I0483798

Deforestation and Climate Change

Ross W. Gorte
Specialist in Natural Resources Policy

Pervaze A. Sheikh
Specialist in Natural Resources Policy

March 24, 2010

Congressional Research Service

7-5700

www.crs.gov

R41144

Summary

Efforts to mitigate climate change have focused on reducing carbon dioxide (CO_2) emissions into the atmosphere. Some of these efforts center on reducing CO_2 emissions from deforestation, since deforestation releases about 17% of all annual anthropogenic greenhouse gas (GHG) emissions and is seen as a relatively low-cost target for emissions reduction. Policies aimed at reducing deforestation are central points of a strategy to decrease carbon emissions, reflected in pending legislation in Congress (e.g., H.R. 2454 and S. 1733) as well as in international discussions, such as the December 2009 negotiations in Copenhagen.

Forests exist at many latitudes. Many are concerned about the possible impacts of losing boreal and temperate forests, but existing data show little, if any, net deforestation, and their loss has relatively modest carbon consequences. In contrast, tropical deforestation is substantial and continuing, and releases large amounts of CO_2, because of the carbon stored in the vegetation and released when tropical forests are cut down.

There are many causes of tropical deforestation—commercial logging, large-scale agriculture (e.g., cattle ranching, soybean production, oil palm plantations), small-scale permanent or shifting (slash-and-burn) agriculture, fuelwood removal, and more. Often, these causes combine to exacerbate deforestation; for example, commercial logging often includes road construction, which in turn opens the forest for subsistence farmers. At times, tropical deforestation results from weak land tenure and/or weak or corrupt governance to protect the forests.

Congress and international bodies are discussing various policies to reduce carbon emissions from deforestation and forest degradation (REDD). Reducing deforestation in the tropics is likely to have additional benefits as well, such as preserving biological diversity and sustaining livelihoods for the rural poor and for indigenous communities and cultures. Proposals may be adapted to address local and regional causes of deforestation. Various forestry practices can reduce the impacts of deforestation, and several market approaches are evolving to compensate landowners for preserving their forests.

Many challenges remain for implementing REDD programs, particularly internationally, including monitoring REDD projects and improving developing-country capacity to ensure compliance. Existing evidence on forests and deforestation suggest the difficulties might be significant. Measuring forests is complicated, with multiple definitions, inaccessible sites, and expensive, complicated, and imperfect measurement technologies.

This report provides basic information on forests and climate change. The first section discusses the linkages between forests and climate. The next three describe the characteristics of the three major forest biomes, with an overview of deforestation causes and impacts. This is followed by an overview of approaches to reducing deforestation. The final section examines issues related to forest and deforestation data.

Contents

Figures

Tables

Appendixes

Contacts

Global climate change is a widespread and growing concern that has led to extensive congressional and international discussions and negotiations. Climate change mitigation strategies have focused on reducing emissions of greenhouse gases (GHGs), especially carbon dioxide (CO_2). One significant source of CO_2 emissions is deforestation. Reducing deforestation to lower CO_2 emissions is seen as one of the least costly methods of mitigating climate change.[1]

Forests are carbon sinks in their natural state (i.e., they store more carbon than they release). Trees absorb CO_2 and convert carbon into leaves, stems, and roots, while releasing oxygen. Forests account for more than a quarter of the land area of the earth, and store more than three-quarters of the carbon in terrestrial plants and nearly 40% of soil carbon. When forests are cleared, some of their carbon is released to the atmosphere—slowly through decay or quickly through burning. One estimate shows that land use change, primarily deforestation, releases about 5.9 $GtCO_2$ (gigatons or billion metric tons of CO_2) annually, about 17% of all annual anthropogenic GHG emissions.[2] This contribution to GHG emissions makes efforts to reduce deforestation significant in international strategies to mitigate climate change.

There has also been some discussion of the relationship between forests and methane (CH_4), a less prominent but far more potent GHG than CO_2. However, the evidence of the relationship is still limited. It generally shows forests to be net CH_4 sinks, except in water-saturated soils (i.e., forested wetlands), and it is unclear whether activities that modify forest cover (e.g., deforestation) affect CH_4 absorption and release.[3] Thus, this report addresses only the relationship between forests and carbon as it affects climate change.

The loss of tropical forests is of particular concern. The existing data show little, if any, net deforestation in boreal and temperate forests, and thus the carbon consequences of deforestation in these ecosystems might not be significant. In contrast, the loss of tropical forests is substantial and continuing. Tropical deforestation has significant climate impacts because of the large amount of CO_2 sequestered in the vegetation—nearly half of all the carbon in terrestrial plants. Thus, the lowest cost and largest carbon benefit of reducing deforestation is with tropical forests. In the United States, tropical forests are largely limited to Hawaii and Puerto Rico.

Congressional Interest

Congress has addressed international deforestation through laws that authorize funding to conserve forests and in proposed climate change legislation that would provide resources to reduce deforestation in developing countries. The Tropical Forest Conservation Act of 1998 (22 U.S.C. §2431 et seq.), for example, authorizes the United States to conduct debt-for-nature swaps

---

[1] See G. Kindermann et al., "Global Costs Estimates of Reducing Carbon Emissions Through Avoided Deforestation," *Proceedings of the National Academy of Sciences*, vol. 105, no. 30 (2008), pp. 10302-10307.

[2] Intergovernmental Panel on Climate Change, "Summary for Policymakers," Climate Change 2007: The Physical Science Basis—Contribution of Working Group I to the Fourth Assessment Report of the Intergovernmental Panel on Climate Change, p. 3, http://www.ipcc-wg1.unibe.ch/publications/wg1-ar4/wg1-ar4.html. The exact amount of CO_2 emissions from deforestation has a large degree of uncertainty, with estimates ranging from 1.8 to 9.9 $GtCO_2$/year for the 1990s. It should be recognized that some sources report gigatons of carbon (C), rather than of CO_2; the conversion is 3.67 tons of CO_2 per ton of carbon.

[3] See, for example, J. P. Megonigal and A. B. Guenther, "Methane Emissions from Upland Forest Soils and Vegetation," *Tree Physiology*, vol. 28, no. 4 (2008), pp. 491-498.

with developing countries to conserve their tropical forests. Under pending climate change legislation (e.g., H.R. 2454, the American Clean Energy Act of 2009, and S. 1733, the Clean Energy Jobs and American Power Act), Congress is considering providing resources for developing countries to establish programs and implement projects to reduce deforestation and forest degradation, and creating policy mechanisms to establish standards and markets for international offsets to reduce GHGs.[4]

Three deforestation issues are likely to be of particular importance to Congress. The first two are geographic variation in the causes and the consequences of deforestation. These then suggest approaches for efforts to reduce deforestation. The third issue is the poor quality of information about forests generally, which might point to needed research and infrastructure as well as suggesting caution in relying on existing data for decision-making.

Congressional interest in reducing deforestation to lower CO_2 emissions parallels several international initiatives that aim to accomplish the same objective. International proposals focus on reducing emissions from deforestation and forest degradation (REDD) in developing countries. These proposals were discussed and debated in climate meetings associated with the United Nations Framework Convention on Climate Change (UNFCCC) in Copenhagen in December 2009.

Forests and Climate

Forest Cover

Forests cover more than a quarter of the land area in the world but are not uniformly distributed. They account for less than 5% of the land in many countries—such as Greenland, Egypt, Pakistan, and Haiti—but cover more than 90% of the land in a few places such as Suriname and French Guiana.[5] Some countries have naturally low forest cover (Greenland and Egypt), whereas others have diminished forest cover because of deforestation, possibly centuries ago (e.g., United Kingdom and Algeria).

Forests store enormous quantities of carbon, and contain more biomass per hectare in vegetation than other biomes. Carbon sequestration and release vary by forest type, although generalizations can be made about the three major forest biomes—boreal, temperate, and tropical forests.[6] **Table 1** shows global average carbon levels in the vegetation and soils for major terrestrial biomes, including the forest biomes. The quantities shown in **Table 1** should be recognized as global averages, with substantial variation of carbon stocks within each biome; for example, wetlands can be dominated by trees (a swamp) or by grasses (a marsh), while tropical forests can be very wet (rainforests) or quite dry (trees in a savannah). There are also continuous gradations across

---

[4] See CRS Report R40990, *International Forestry Issues in Climate Change Bills: Comparison of Provisions of S. 1733 and H.R. 2454*, by Pervaze A. Sheikh and Ross W. Gorte.

[5] Food and Agriculture Organization of the United Nations (FAO), *Global Forest Resources Assessment 2005*, FAO Forestry Paper 147, Rome, Italy, 2006. Hereafter cited as FAO 2005 FRA. The report and tables on forest area are at http://www.fao.org/forestry/fra2005/en/.

[6] A *biome* is defined as a "regional land-based ecosystem type ... characterized by consistent plant forms and ... found over a large climatic area." Henry W. Art, ed., *The Dictionary of Ecology and Environmental Science* (New York, NY: Henry Holt & Co., 1993), p. 65.

---

biomes (e.g., warm, humid temperate forests—subtropical forests—have traits in common with both temperate forests and tropical forests).

Table 1. Average Carbon Stocks for Various Biomes

(area in billion hectares; carbon in metric tons of CO_2 per hectare)

Biome	Area	Plant Carbon	Soil Carbon	Total Carbon
Tropical forests	1.76	442	450	892
Temperate forests	1.04	208	352	561
Boreal forests	1.37	236	1,260	1,496
Tundra	0.95	23	467	490
Croplands	1.60	7	293	300
Tropical savannas	2.25	108	430	538
Temperate grasslands	1.25	26	865	892
Deserts/semi-desert lands	4.55	6	154	160
Wetlands	0.35	157	2,357	2,514
Weighted average across total area	*15.12*	*113*	*488*	*601*

Source: Adapted from Intergovernmental Panel on Climate Change, "Table 1: Global Carbon Stocks in Vegetation and Carbon Pools Down to a Depth of 1 m [meter]," *IPCC Special Report on Land Use, Land-Use Change and Forestry: Summary for Policymakers* (2000), http://www.ipcc.ch/ipccreports/sres/land_use/003.htm.

Notes: Area column sum may not match reported total because of rounding error. Includes only CO_2, and not other GHGs, such as CH_4 and N_2O.

Linkages Between Forests and Climate

Deforestation is the loss of tree cover, usually as a result of forests being cleared for other land uses such as farming or ranching. Some limit the definition of deforestation to the permanent conversion of forests to another habitat. Others add to this definition by including the conversion of natural forests to artificial forests such as plantations.[7]

Deforestation activities affect carbon fluxes in the soil, vegetation, and atmosphere.[8] The effects of these activities can vary, depending on the type of activity. For example, logging can lead to carbon storage if trees are converted to wood products (e.g., lumber) and deforested areas are restored.[9] (The issues surrounding tree planting to offset deforestation are discussed below.)

---

[7] Forest *degradation* is where some, but not all, of the tree cover is removed or destroyed. This results from activities such as logging or fuelwood extraction that remove specific trees, usually because of their commercial wood value. The remaining forest generally has reduced canopy cover, and remaining trees often suffer collateral damage, either from the removal process or from roads. Degraded forests also often have lower species diversity and reduced regeneration of seedlings and saplings. Many of the effects of forest degradation are similar to the effects of deforestation, and efforts to reduce deforestation often include efforts to reduce forest degradation—the goal is known as "reduced emissions from deforestation and degradation," or REDD. However, this report focuses exclusively on deforestation.

[8] For information on carbon fluxes—carbon releases and sequestration—see CRS Report RL34059, *The Carbon Cycle: Implications for Climate Change and Congress*, by Peter Folger.

[9] The proportion of carbon stored in products varies widely, depending on the species, size, and form of the trees cut, the nature of the harvesting, and the products made. For more information on forest carbon accumulation and release, (continued...)

However, logging can also lead to carbon emissions if the surrounding trees and vegetation are damaged, and if not all woody biomass is processed into products. Other activities that alter the carbon cycle in forests and affect climate are discussed below.

Soil Impacts

The impact of deforestation on soils, and the release of soil carbon, depends on the magnitude of soil disturbance and the type of soil. Soil carbon content is related to the lifecycle of the vegetation it supports. When vegetation dies, it decomposes and releases carbon. Some carbon is deposited in the soil; some is dissolved and leaches into surface waters or groundwater; and some is released directly to the atmosphere as CO_2. Deforestation exposes soils to sunlight, which increases soil temperature and the rate of soil carbon oxidation. This process increases the rate of CO_2 release to the atmosphere. Soil carbon can also be released at high rates if soils are disturbed, for example, by logging operations or tillage.

Peat soils are particularly important for climate because of their very high carbon content (as well as CH_4 content and release). Peat soils generally occur in forests where natural decomposition rates are low, such as in periodically flooded forests or forests with a short growing (and thus decomposition) season. Peat soils may partly account for the high soil carbon levels of boreal forests, but some occur in temperate and tropical forests. Peat soils are considered major carbon sinks, and could potentially be large sources of carbon emissions, if disturbed.

Wood Utilization/Wood Waste

Deforestation can lead to carbon emissions from decomposing vegetation left on the forest floor. However, wood converted into products—such as lumber and plywood—could store carbon for many years, ranging from an average of 10 years for shipping pallets to 100 years or more for lumber.[10] Paper products store carbon for a brief duration, often less than a year. The proportion of a tree converted to products varies widely, and depends on the size (diameter) and form (taper and branching) of the tree as well as the particular species. The purpose of the tree cutting also affects utilization and waste. Harvesting pulpwood for paper production, for example, can include much more of the woody biomass than harvesting veneer bolts for plywood or sawlogs for lumber. Cutting to clear a site for agriculture yields much more waste, as the woody biomass is generally burned to prepare the site for crop or pasture production.

In addition, the harvest method can affect wood utilization and waste. Selective logging, where certain trees or species are harvested, can lead to large quantities of wood waste because more roads are needed and because the harvest and extraction procedures often damage the remaining trees. Clear-cutting can reduce wood waste, when the majority of the trees can be removed for wood products, but can increase biomass waste if done to clear land for agriculture. One technique—reduced impact logging (RIL)—has been developed to reduce timber harvest damage

---

(...continued)

see CRS Report RL31432, *Carbon Sequestration in Forests*, by Ross W. Gorte.

[10] K. E. Skog and G. A. Nicholson, "Carbon Sequestration in Wood and Paper Products," *The Impact of Climate Change on America's Forests: A Technical Document Supporting the 2001 USDA Forest Service RPA Assessment*, USDA Forest Service, Rocky Mountain Research Station, Gen. Tech. Rept. RMRS-GTR-59, Ft. Collins, CO, 2000, pp. 79-88.

---

to soils and residual trees.[11] Descriptions of RIL are typically either lacking in details or highly site-specific with limited general applicability, because the practices that will reduce logging damages depend on a variety of site conditions, such as soil type and water content, and tree species diversity. Nonetheless, one source reported that RIL reduces wood waste by more than 60% and soil disturbance in roads, landings, and skid trails by almost 50%.[12] However, a major barrier to increased use of RIL is illegal logging in the tropics.[13]

Biomass not removed for products remains on the site and decomposes. Some of the carbon is deposited in the soil and some is released into the atmosphere. If the remaining biomass is burned, as is common in clearing lands for agriculture and in preparing sites for reforestation, the carbon is quickly released to the atmosphere. For unburned on-site biomass, the rate of decomposition (and hence of carbon emissions) varies due to moisture (many fungi and bacteria grow better in moist environments), temperature (higher temperatures also improve fungi and bacteria growth), and type of wood (some species contain chemicals that inhibit decomposers), among other things.

Burning—Natural and Anthropogenic

Forest fires—both natural and anthropogenic—can kill some or all of the trees in a forest. Forested ecosystems have evolved with a variety of natural fire regimes. Some ecosystems have rare natural fires; others are "fire-prone."[14] The nature and extent of natural fires are related to the evolutionary development of the natural fire regimes, to climatic conditions such as drought, and to the amount of woody fuels in some ecosystems.[15] Fire affects climate by releasing large quantities of CO_2 to the atmosphere in short periods, and thus extensive burning can affect the global climate.[16] Fires also produce large quantities of fine particulates and aerosols. These pollutants can be hazardous to human health, but they also absorb and reflect sunlight, which creates cooler temperatures in the forest.[17]

Anthropogenic burning is a greater concern for carbon emissions than natural fires. For example, fire is commonly used to clear land in the tropics. Studies report that anthropogenic ignitions are the predominant factor in starting wildfires in tropical forests.[18] Man-made fires in areas not prone to natural fires can lead to a positive feedback loop, where increasing fire frequency can alter plant regrowth until forests do not regenerate and the areas are converted to brush fields. In contrast, natural fires are part of the carbon cycle, with carbon emissions balanced by plant

---

[11] See D. P. Dykstra, *Reduced Impact Logging: Concepts and Issues*, FAO Corporate Document Repository, http://www.fao.org/docrep/005/ac805e/ac805e04.htm.

[12] Tropical Forest Foundation, "Reduced Impact Logging," at http://www.tropicalforestfoundation.org/get-verified/reduced-impact-logging.

[13] See CRS Report RL33932, *Illegal Logging: Background and Issues*, by Pervaze A. Sheikh.

[14] M. A. Krawchuk et al., "Global Pyrogeography: The Current and Future Distribution of Wildfire," *PLoS One*, vol. 4, no. 4 (April 2009), pp. 1-12.

[15] For more information, see CRS Report R40811, *Wildfire Fuels and Fuel Reduction*, by Ross W. Gorte.

[16] T. M. Bonnicksen, *Greenhouse Gas Emissions from Four California Wildfires: Opportunities to Prevent and Reverse Environmental and Climate Impacts*, The Forest Foundation, FCEM Report No. 2, Auburn, CA, March 12, 2008, pp. 1-19.

[17] V. Ramanathan et al., "Warming Trends in Asia Amplified by Brown Cloud Solar Absorption," *Nature*, vol. 448 (2007), pp. 575-578.

[18] M. A. Cochrane and C. P. Barber, "Climate Change, Human Land Use and Future Fires in the Amazon," *Global Change Biology*, vol. 15, no. 3 (November 2008), pp. 601-612.

regrowth over the long run.[19] According to some, this balance justifies not controlling natural fires to mitigate climate change.[20] Others contend that natural fires could exacerbate the ecological effects of anthropogenic fires on forest ecosystems, and argue that both should be regulated and controlled, depending on the location and fire history of the site.[21]

Other Relationships Between Deforestation and Climate Change

Four other relationships between deforestation and climate change are discussed in the literature:

- **Disturbances Other Than Fire.** Forests are disturbed by insect infestations, disease, drought, invasive species, wind and ice storms, and landslides, among other things. Understanding of how these disturbances relate to climate change is generally incomplete.[22] An exception is the relationship between increased pine beetle infestations in the Rocky Mountains and warmer temperatures. Warmer temperatures allow pine beetles to increase their seasonal reproductive rate and expand their range among pine stands.[23] Some are concerned that climate change will exacerbate forest disturbances, leading to further climate change (i.e., creating a positive feedback loop).[24]

- **Albedo Effect.** Albedo is a measure of the reflectivity of surfaces (e.g., vegetation, soils, and water)—darker surfaces absorb more sunlight (e.g., fir forests), while light-colored surfaces reflect more sunlight (ice or snow). Darker surfaces heat the surrounding atmosphere more than lighter surfaces, making albedo important for climate. Climate models have shown that the reduced surface heating from the very high albedo of snow-covered boreal forest openings more than offsets the warming from the CO_2 released in creating those openings.[25] For temperate forests, where the openings are snow-covered for a briefer period, the albedo effect is relatively minor. For tropical forests, where there is no snow and where the vegetation is a broader mix of species, there is no significant difference in albedo between the forest canopy and forest openings.

- **Carbon Dioxide Fertilization of Forests.** Because CO_2 is critical to vegetative growth, some have hypothesized that elevated CO_2 levels will increase forest growth. Studies of temperate forests report that excess carbon in the atmosphere can increase photosynthesis and plant growth in trees for short periods, particularly for young stands.[26] Long-term effects are unknown because

---

[19] D. M. J. S. Bowman et al., "Fire in the Earth System," *Science*, vol. 324 (April 24, 2009), pp. 481-484.

[20] The Wilderness Society, *Climate Change Facts: Fossil Fuels Are a Bigger Problem than Wildland Fires*, February 7, 2008, http://wilderness.org/files/Wildland-Fire-Fossil-Fuels.pdf.

[21] M. D. Hurteau, G. W. Koch, and B. A. Hungate, "Carbon Protection and Fire Risk Reduction: Toward a Full Accounting of Forest Carbon Offsets," *Frontiers in Ecology and the Environment*, vol. 6 (2008).

[22] V. H. Dale et al., "Climate Change and Forest Disturbances," *BioScience*, vol. 51, no. 9 (September 2001), pp. 723-734.

[23] For more information, see CRS Report R40203, *Mountain Pine Beetles and Forest Destruction: Effects, Responses, and Relationship to Climate Change*, by Ross W. Gorte.

[24] See, for example, J.A. Logan, J. Règniére, and J.A. Powell, "Assessing the Impacts of Global Warming on Forest Pest Dynamics," *Frontiers in Ecology and the Environment*, vol. 1, no. 3 (2003): pp. 130-137.

[25] See G. B. Bonan, "Forests and Climate Change: Forcings, Feedbacks, and the Climate Benefits of Forests," *Science*, vol. 320 (June 13, 2008), pp. 1444-1449.

[26] See Duke University, "Experiment Suggests Limitations to Carbon Dioxide 'Tree Banking,'" News & Communications, Durham, NC, August 27, 2007, http://news.duke.edu/2007/08/carbonadd._print.ht.

---

experiments need more time to collect data. Other researchers note that sustained enhanced growth due to high CO_2 levels may be limited by factors such as drought and nitrogen availability.[27] For tropical forests, some suggest that carbon saturation by leaves of tropical trees may limit response to CO_2 enrichment, and decreased productivity could result from periods of higher temperatures and drought.[28]

- **Hydrological Patterns.** Climate change can directly alter precipitation patterns, sometimes causing drought in some areas. Researchers report that higher CO_2 levels and temperatures increase water use by plants.[29] The combination of drought and demand for greater water could stress forests and cause changes in the ecosystem. In contrast, broad-scale deforestation has been shown to reduce evapotranspiration (water loss to the atmosphere) by plants, which reduces cloud formation and downwind precipitation.[30] The combination of changes in precipitation patterns, plant water use, and evapotranspiration could have significant synergistic effects.

Boreal Forests

Boreal forests, or *taiga*, generally occur north of about 50° north latitude, as shown in **Figure 1**. Although boreal forests account for about a third of the world's forests (see **Table 1**), relatively few countries have boreal forests. Countries with boreal forests include Russia, Canada, the United States (in Alaska), Sweden, Finland, and Norway.[31] There are few boreal forests in the Southern Hemisphere, including minor acreages on scattered mountaintops in southern Argentina, Chile, and New Zealand (not shown in **Figure 1**).

Boreal forests are dominated by relatively few tree species, such as spruce, fir, larch, and pine. They commonly grow in expanses of trees with relatively similar sizes, generally as a result of infrequent broad-scale destructive events, particularly wildfires. These conifer forests often contain substantial volumes of timber, but they generally are not managed for timber production, because of slow growth rates. Boreal forests are important for carbon sequestration because of their high carbon storage in forest soils. (See **Table 1**.) Carbon in vegetation is slightly greater than for temperate forests, and about half of the level in tropical forests. However, soil carbon levels in boreal forests are high—more than for any other biome except wetlands. Carbon accumulates to high levels in boreal soils because of slow decomposition rates, which are depressed by short summers and acidic soils.

---

[27] S. V. Ollinger et al., "Canopy Nitrogen, Carbon Assimilation, and Albedo in Temperate and Boreal Forests: Functional Relations and Potential Climate Feedbacks," *Proceedings of the National Academy of Sciences*, vol. 105, no. 49 (December 9, 2008), pp. 19336-19341.

[28] D. A. Clark, "Sources or Sinks? The Responses of Tropical Forests to Current and Future Climate and Atmospheric Composition," *Philosophical Transactions of the Royal Society of Biological Sciences*, vol. 359 (2004), pp. 477-491.

[29] See, for example, R. L. Graham, M. G. Turner, and V. H. Dale, "How Increasing CO_2 and Climate Change Affect Forests," *BioScience*, vol. 40, no. 8 (September 1990), pp. 575-587.

[30] A. J. Hansen et al., "Global Change in Forests: Responses of Species, Communities, and Biomes," *BioScience*, vol. 51, no. 9 (September 2001), pp. 765-779.

[31] These countries also contain temperate forests in their southern reaches, but generally contain much greater forestlands and timber stands in boreal forests.

Figure 1. Northern Hemisphere Boreal Forests

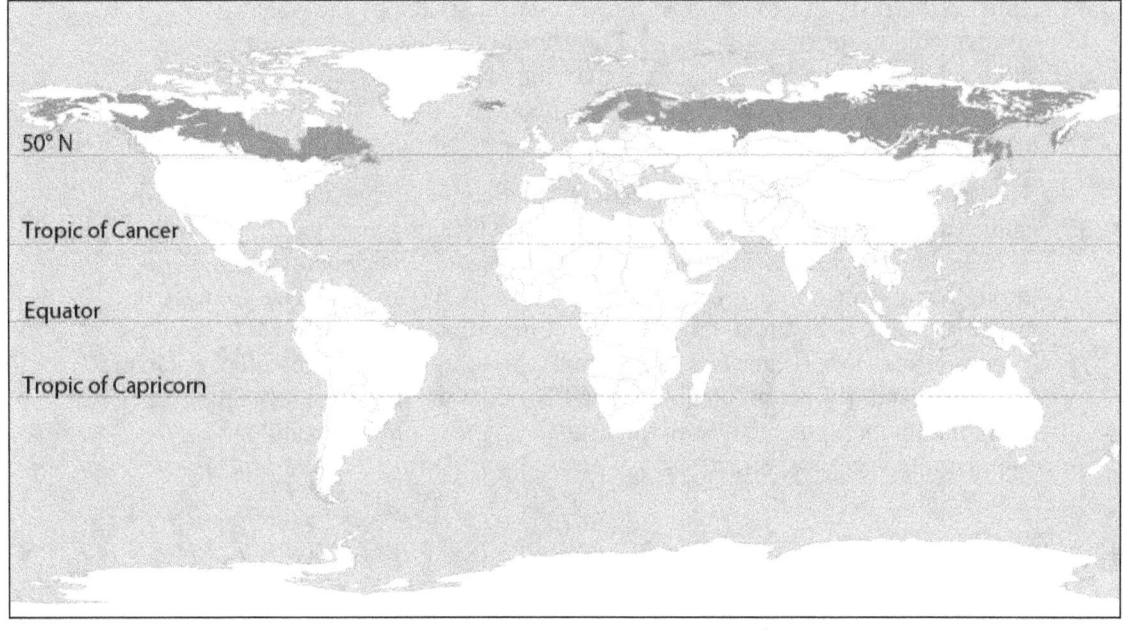

Source: Prepared by CRS based on World Wildlife Fund, Terrestrial Ecosystem, http://www.worldwildlife.org/science/data/item1875.html. Original source is D. M. Olson et al., "Terrestrial Ecosystems of the World: A New Map of Life on Earth," *Bioscience*, vol. 51 (2001), pp. 933-938.

Causes of Boreal Deforestation

The primary driver of boreal deforestation has historically been land clearing for agriculture, primarily along the southern borders of boreal forests.[32] It is unclear whether forest clearing for agriculture is continuing, although warming from global climate change might make some boreal ecosystems ideal for some types of agriculture. Another possible cause of boreal deforestation is timber harvesting. Some contend that logging is a significant driver of deforestation in boreal forests;[33] others suggest that logging is secondary to the effects of increased wildfires and insect and disease infestations.[34] Boreal forests are not typically managed intensively for timber production, but the substantial volumes of standing timber are harvested extensively for wood products in some regions such as Scandinavia, western Russia, and parts of Canada.

Limited evidence of continuing permanent deforestation in boreal forests is inconclusive. Some research has found virtually no regeneration of the boreal forests in eastern Canada following wildfires over the past 900 years, with a commensurate decline in forest cover.[35] These findings are contrary to the "widespread belief of northward expansion of forests due to recent warming"

---

[32] K. A. Hobson, E. M. Bayne, and S. L. Van Wilgenburg, "Large-Scale Conversion of Forest to Agriculture in the Boreal Plains of Saskatchewan," *Conservation Biology*, vol. 16, no. 6 (December 2, 2002), pp. 1530-1541.

[33] See, for example, P. Janes, "Vanishing Forest: A Northern Forest Is Disappearing at a Rapid Pace—That Spells Trouble for Billions of Animals," *Science World*, March 27, 2006.

[34] K. Jardine, *The Carbon Bomb: Climate Change and the Fate of the Northern Boreal Forests*, Greenpeace International, Amsterdam, The Netherlands, 1994, http://dieoff.org/page129.htm.

[35] S. Payette, L. Filion, and A. Delwaide, "Spatially Explicit Fire-Climate History of the Boreal Forest-Tundra (Eastern Canada) Over the Last 2000 Years," *Philosophical Transactions of the Royal Society of Biological Sciences*, vol. 363, no. 1501 (November 28, 2007), pp. 2301-2316.

---

and suggest that boreal forests might not migrate northward in response to climate change, as many believe.[36] Others have noted "enhanced conifer recruitment" (forest regeneration) in Russian boreal forests in the 20th century.[37] Thus, it is difficult to draw conclusions about the extent of deforestation in boreal forests.

Climate Consequences of Possible Boreal Deforestation

The loss of boreal forests could have several possible consequences for climate. When boreal forest trees are lost—through wildfire, insect or disease infestation, or timber harvest—at least some of the carbon contained in the trees and soils is emitted to the atmosphere. The amount of carbon released depends on many factors. However, because about five-sixths of boreal forest carbon is stored in the soil, soil disturbance is the most important factor for carbon release. The loss of forest cover from deforestation may accelerate the oxidation of carbon near the soil surface and cause increased emissions. Timber harvesting and the associated road construction disturb soils, and can release substantial amounts of soil carbon to the atmosphere. On the other hand, winter harvesting over packed snow can significantly reduce soil disturbance, although it is more costly than traditional timber harvesting.

Another factor that reduces the climate impacts of boreal deforestation is the high level of wood utilization and relatively low wood waste. Because of the relatively low tree species diversity of boreal forests, clearcutting is common and most trees can be used for wood products, including those killed by wildfire or by insects and diseases, both of which reduce wood waste. Furthermore, pulpwood harvesting for paper production (short-term wood products) is especially common in boreal forests, also leading to less wood waste in the forest to decompose and release carbon.

Another possible climate impact of boreal deforestation is related to the albedo effect. The dominant species in boreal forests—firs, spruces, larches, and pines—are relatively dark-colored, and thus absorb much of the incoming sunlight. When these species are cleared, different species take their places, primarily broadleaf species such as aspen, alder, and birch. These trees are lighter in color and reflect much of the incoming sunlight. Furthermore, snow accumulates in clearings, and reflects more sunlight than snow under the trees. Thus, boreal forest clearings are cooler than the surrounding forest, which could create a local cooling effect and slow decomposition. Some models have suggested that the cooling from the albedo effect more than offsets the warming from the carbon released, even from wildfires.[38] Other researchers, however, have calculated that the increased transpiration of light-colored broadleaf forests as they displace tundra more than offsets the albedo effect, resulting in additional warming.[39]

---

[36] Payette, Filion, and Delwaide, "Spatially Explicit Fire-Climate History."

[37] G. M. MacDonald, K. V. Kremenetski, and D. W. Beilman, "Climate Change and the Northern Russian Treeline Zone," *Philosophical Transactions of the Royal Society of Biological Sciences* vol. 363, no. 1501 (November 15, 2007), pp. 2285-2299.

[38] G. Bala et al., "Combined Climate and Carbon-Cycle Effects of Large-Scale Deforestation," *Proceedings of the National Academy of Sciences*, vol. 104, no. 16 (April 17, 2007), pp. 6550-6555.

[39] A. L. Swann et al., "Changes in Arctic Vegetation Amplify High-Latitude Warming Through the Greenhouse Effect," *Proceedings of the National Academy of Sciences*, vol. 107, no. 4 (January 2010), pp. 1295-1300.

Temperate Forests

Temperate forests generally occur in the mid-latitudes, typically from the Tropic of Cancer (23½° north latitude) to about 50° north latitude, and south of the Tropic of Capricorn (23½° south latitude), as shown in **Figure 2**. Temperate forests account for about a quarter of global forests. The most extensive temperate forests are in the United States and southern Canada, Europe, China, and Australia.

Figure 2. Temperate Forests of the World

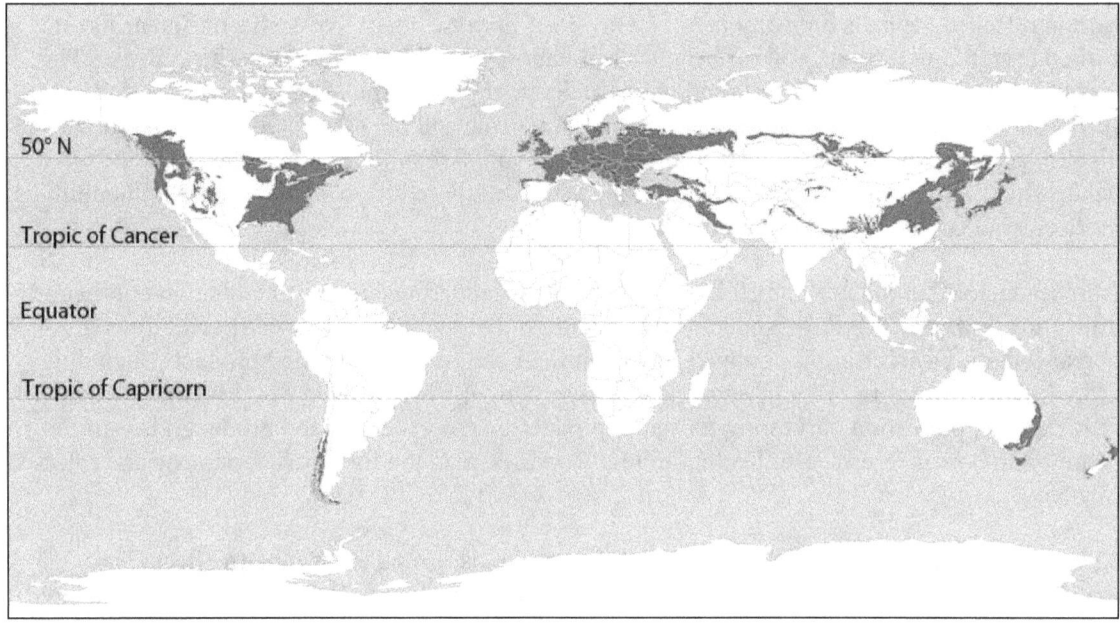

Source: Prepared by CRS based on World Wildlife Fund, Terrestrial Ecosystem, http://www.worldwildlife.org/science/data/item1875.html. Original source is D. M. Olson et al., "Terrestrial Ecosystems of the World: A New Map of Life on Earth," *BioScience*, vol. 51 (2001), pp. 933-938.

There is a wide variety of temperate forests—oak, maple, pine, and more—but the species diversity within temperate forests, while greater than in boreal forests, is substantially lower than in tropical forests. As with boreal forests, temperate forests commonly have extensive areas covered by a few tree species with similar sizes, often the result of destructive events—wildfires and major storms (hurricanes, tornadoes, wind or ice storms, etc.). Many temperate forests are managed for commercial wood production, because of the modest species diversity, moderate tree growth rates, and desirable wood characteristics of many of the dominant conifer trees. Temperate forests are less significant for carbon release than tropical or boreal forests, because of the lower levels of carbon stored in vegetation and soils. However, they take on added significance because of the more intensive management of these forests for wood products.

Causes of Temperate Deforestation

The disturbances in temperate forests parallel those of boreal forests. Clearing forests for agriculture is a historical cause of deforestation in many areas, although much of this occurred more than a century ago. Clearing land for agriculture might still be a cause of deforestation in temperate developing countries.

A more common cause of temperate deforestation, particularly in developed countries, is conversion of land to non-agricultural uses, notably residential and commercial development and infrastructure (e.g., roads).[40] While research has suggested that no net loss of temperate U.S. forests is anticipated, changes in forest cover are likely to fragment remaining forestland.[41] The implications of such changes for carbon sequestration and climate are unclear.

Timber harvesting and natural disasters are also causes of temperate deforestation. In many areas, timber harvesting is followed by reforestation, resulting in no net deforestation over time. Similarly, most areas affected by natural disasters are usually reforested, either naturally or through tree planting. Nonetheless, reforestation of cleared areas is not always successful, because of drought and invasion by competing species (native brush, plants used for erosion control, invasive exotics, etc.). In some areas, natural succession may require years or decades to reestablish tree cover, and climate change may prevent such "normal" recovery.

Climate Consequences of Possible Temperate Deforestation

Deforestation of temperate forests generally has less severe consequences for climate than tropical or boreal deforestation because of the lower carbon levels in soils and vegetation. However, insect outbreaks and fires in Canadian temperate forests have transformed these forests since 2000 from a carbon sink to a projected carbon source, a status expected to continue for the next two to three decades.[42] In the western United States, historic fire suppression policies have increased biomass, but have also increased the risk of catastrophic fires. More recently, policies have focused on reducing biomass fuels, through prescribed burning and mechanical treatments. These activities release carbon, but increase tree growth (and carbon sequestration) and reduce CO_2 released from wildfires in some areas.[43] Thus, the net effects of wildfires and of efforts to reduce wildfire damages are unclear.

Disturbance of temperate soils can be a factor in carbon emissions depending on its intensity. Natural disasters typically do not disturb the soils, although the loss of forest cover may accelerate erosion and the oxidation of soil carbon. Timber harvesting and its associated road construction disturb soils and can release substantial amounts of soil carbon. Logging could also affect carbon emissions from temperate forests directly. Large-scale clearing, followed by burning and decomposition, could increase carbon emissions; however, wood utilization from timber harvesting (including salvage of trees killed by fire, insects, and diseases) will increase carbon storage, and reforestation provides additional carbon sequestration.[44] The net effect is unclear. (See "Tree Planting to Offset Deforestation?" below.)

---

[40] This is a concern for many forests in the southern United States. See S. M. Stein et al., *Forests on the Edge: Housing Development on America's Private Forests*, USDA Forest Service, Pacific Northwest Research Station, Gen. Tech. Rept. GTR-PNW-636, Portland, OR, 2005.

[41] A. J. Plantinga et al., *Linking Land-Use Projections and Forest Fragmentation Analysis*, USDA Forest Service, Pacific Northwest Research Station, Res. Pap. PNW-RP-570, Portland, OR, 2007.

[42] W. A. Kurz et al., "Risk of Natural Disturbances Makes Future Contribution of Canada's Forests to the Global Carbon Cycle Highly Uncertain," *Proceedings of the National Academy of Sciences*, vol. 105, no. 5 (February 2008), pp. 1551-1555.

[43] M.D. Hurteau and M. North, "Fuel Treatment Effects on Tree-Based Forest Carbon Storage and Emissions Under Modeled Wildfire Scenarios," *Frontiers in Ecology and the Environment*, vol. 7 (2009).

[44] See CRS Report RL31432, *Carbon Sequestration in Forests*, by Ross W. Gorte.

Tropical Forests

Tropical forests are generally defined by their location—between the Tropic of Cancer and the Tropic of Capricorn, 23½° north and south of the equator, respectively. Tropical forests occur in many settings, from very wet to quite dry locations. Tropical rainforests, shown in **Figure 3**, are the dominant form, characterized by heavy rainfall, dense vegetation, and an enormous diversity of plant and animal species. Tropical rainforests are considered to be among the earth's most biologically diverse ecosystems; indeed, some claim that tropical rainforests hold nearly 50% of the earth's biodiversity. Dry tropical forests have sparser tree cover and less species variability, typically with grasses and other herbaceous vegetation growing underneath.

Figure 3. Tropical Rainforests of the World

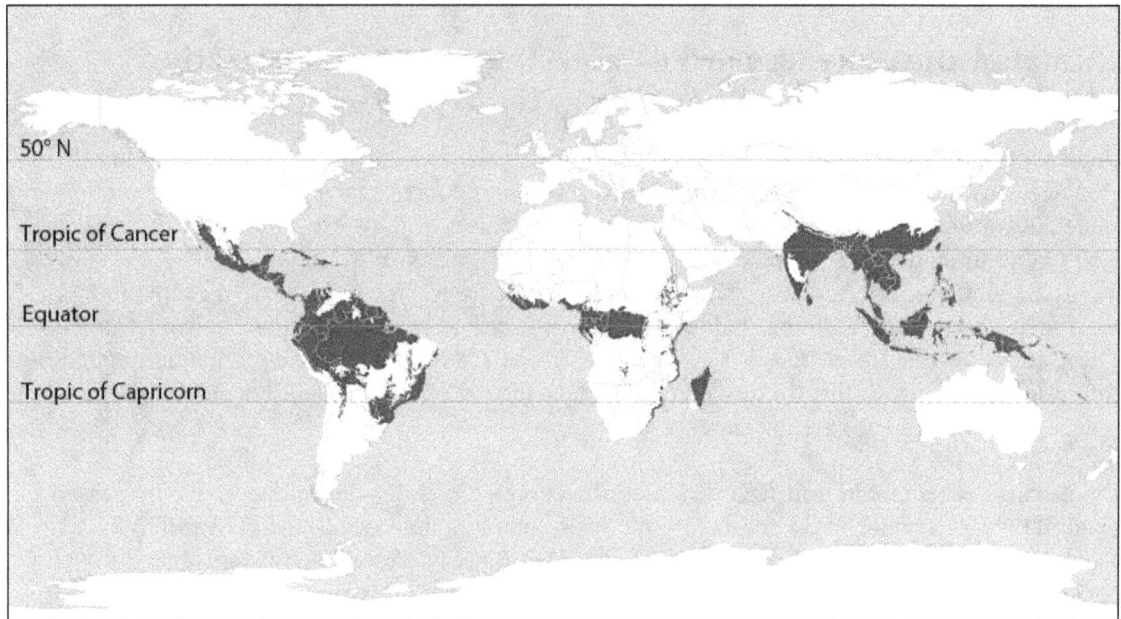

Source: Prepared by CRS based on World Wildlife Fund, Terrestrial Ecosystem, http://www.worldwildlife.org/ science/data/item1875.html. Original source is D. M. Olson et al., "Terrestrial Ecosystems of the World: A New Map of Life on Earth," *Bioscience*, vol. 51 (2001), pp. 933-938.

Tropical countries account for about 42% of global forestlands. Rainforests are common in Central and South America, southern and Southeast Asia, and the Congo Basin and Madagascar in Africa. The countries with the most extensive tropical rainforests include Brazil, the Democratic Republic of Congo, and Indonesia. About half of dry tropical forests occur in a band from the easternmost part of Brazil southwest into Paraguay and eastern Bolivia, and scattered elsewhere in Latin America (e.g., in Mexico). Other dry tropical forests occur in a band across Africa, south of the Sahara Desert, southward through East Africa, and south of the Congo Basin, as well as scattered in southern Asia and in northern Australia.

Tropical forests have an enormous diversity of plant and animal species. In contrast to boreal and temperate forests, tropical forests generally have been free from infrequent, broad-scale destructive events. Thus, the trees (and other species) on each site can respond to minor localized climatic differences that, over thousands of years, can lead to diversification. As a result, tropical areas are generally not well suited for intensive forest management or plantations, although teak and mahogany (as well as coffee, oil palms, and bananas) are sometimes grown in plantations.

Many of the desired species have narrow habitat requirements, often making it difficult for them to grow near other trees of the same species or requiring a variety of species to provide the necessary micro-climatic conditions. Further, modest soil carbon levels and rapid decomposition effectively prevent sustained intensive management over extended periods without substantial and continuing applications of fertilizers. The wide variety of trees also leads to a wide variety of insects and diseases, so pest management is an issue for tropical plantations.

Tropical forests are important for carbon sequestration. They contain substantial amounts of carbon in vegetation—double the level in other forests, and four times more carbon than the global average. (See **Table 1**, above.) In contrast to the vegetative carbon, tropical forest soils contain only average levels of carbon. In tropical rainforests, the carbon is quickly depleted when vegetation is cut, because the warm, humid conditions cause rapid decomposition and the high rainfall leaches minerals from the soils.

Causes of Tropical Deforestation

Compared to boreal and temperate deforestation, tropical deforestation is expected to have more significant climate consequences because of the higher rate and amount of CO_2 released. Policy mechanisms that address deforestation are related to drivers of deforestation. In the tropics, drivers of deforestation vary among regions, and thus a single solution for reducing deforestation in the tropics might be insufficient.

This section discusses some common anthropogenic drivers of tropical deforestation. There are direct anthropogenic drivers of tropical deforestation, such as clearing for agriculture, as well as underlying causes, such as road construction, market forces, and government policies.[45] Underlying drivers of deforestation are generally coupled with direct drivers of deforestation. Tropical forest losses from anthropogenic causes can be exacerbated by natural events, such as drought and fire, as discussed earlier.[46]

Some have identified the drivers of tropical deforestation in several categories.[47] There are direct drivers—agriculture, including shifting cultivation and small-scale and large-scale permanent agriculture; and wood extraction, including logging and fuelwood harvests. In addition, there are two principal underlying causes of deforestation—road building and governmental policies.

The anthropogenic drivers of deforestation vary among regions, and differ between rainforests and dry tropical forests. Drivers of deforestation seldom act independently of other drivers and in many cases follow a progression.[48] For example, in some regions of the Amazon Basin, selective logging for high-value timber species leads to road building through virgin tracts of forests. After logging, farmers use logging roads to access forested areas. Farmers cut trees and burn them to prepare areas for planting crops or forage for ranching. Another example is the coupling of

---

[45] Helmut J. Geist and Eric F. Lambin, *What Drives Tropical Deforestation?* (Louvain, Belgium: LUCC International Project Office, 2001), pp. 5-22.

[46] L. E. O. C. Aragão et al., "Interactions Between Rainfall, Deforestation and Fires During Recent Years in the Brazilian Amazon," *Philosophical Transactions of the Royal Society of Biological Sciences*, vol. 363, no. 1498 (May 2008), pp. 1779-1785.

[47] See, for example, Michael Williams, *Deforesting the Earth: From Prehistory to Global Crisis, An Abridgment* (Chicago, IL: University of Chicago Press, 2003), p. 459.

[48] Geist and Lambin, *What Drives Tropical Deforestation?*

deforestation with population increases, a sequence common for tropical regions in Asia and Africa. As population levels rise, demand for agricultural products increases. This demand can lead to expanding agricultural fields into areas previously occupied by forests.

Road Building

Building roads into forested areas is the major underlying cause of deforestation in the tropics. Road building increases access to forested land and is the first step toward developing forested regions, often for agriculture.[49] Thus roads often initiate the deforestation process.[50] Two scales of road building commonly occur. Roads may be built on small scales into virgin forests, typically to access trees for selective logging. Roads are also built by governments on large scales to connect regions within a country. Newly built roads provide access to forested areas to initiate logging, agriculture, and ranching, and provide return access to markets to sell products. Roads can also stimulate development, leading to infrastructure and market development at the forest frontier. They make transportation cheaper and can encourage migration to forest frontiers, which increases pressure on forests.[51] In many tropical areas, the ecological and economic consequences and construction methods of road-building are rarely considered in planning.

Agriculture

Agriculture in the tropics is diverse, but is traditionally associated with shifting cultivation, also known as *slash-and-burn* farming or swidden agriculture. This practice involves clearing a site by cutting down and burning trees, growing crops until soils are depleted of nutrients (a few years), and then moving to a new site and repeating the process. The process is associated with about a third of agriculture-related deforestation throughout the tropics.[52]

Shifting cultivation is being displaced by small-scale permanent agriculture, where the same site is farmed indefinitely. As with shifting cultivation, small-scale agriculture begins with cutting and burning the trees. Permanent small-scale agriculture in the tropics requires greater investment in the land (e.g., use of fertilizer for fortifying soils) and increases land tenure, thereby slowing the rate of deforestation after initially contributing to deforestation. Several factors enter the decision-making process when farmers consider whether to practice swidden agriculture or permanent agriculture, including the cost of maintaining a plot, the cost of acquiring new land, access to roads and infrastructure, and land rights. Permanent agriculture is a driver of deforestation as the number and size of plots expand due to increased demand for food and biofuels. Permanent small-scale agriculture is associated with about another third of agriculture-related tropical deforestation.[53]

Large-scale permanent agriculture is a third variety of tropical agriculture associated with deforestation. Some of these operations involve traditional crops such as bananas and coffee;

---

[49] T. Tomich, ed., *Forces Driving Tropical Deforestation*, ASB (Alternatives to Slash-and-Burn) PolicyBriefs 06, Nairobi, Kenya, November 2003.

[50] J. Eliasch, *Climate Change: Financing Global Forests, the Eliasch Review* (London: Earthscan, 2008), pp. 47-48.

[51] Eliasch, *Climate Change: Financing Global Forests*, p. 43.

[52] Geist and Lambin, *What Drives Tropical Deforestation?* p. 25. Geist and Lambin report multiple, interactive causes of deforestation, and thus their total of all causes of deforestation substantially exceeds 100%.

[53] Geist and Lambin, *What Drives Tropical Deforestation?* p. 25.

others involve crops such as soybeans and oil palm that meet the demands of emerging markets (e.g., biofuels). Large-scale permanent agriculture also covers large-scale ranching operations and forest plantations. Large-scale agriculture and cattle ranching roughly account for the remaining third of agriculture-related deforestation in tropical countries.[54]

Wood Extraction

Logging and fuelwood consumption are primary causes of tropical deforestation associated with wood extraction. Commercial logging in the tropics is largely selective. Tropical forests have a large diversity of tree species per acre, of which a few are valuable in commercial markets. Logging crews often create roads and trails through forests to access select valuable species (i.e., selective logging), often leaving cut and damaged trees. For a few particular species, the unused woody biomass can have more mass than the timber collected. While logging might not be a significant direct cause of deforestation, it can be a significant underlying cause. As discussed above, logging often opens forests to agricultural expansion.

Reduced impact logging (RIL) can reduce timber harvest damage to tropical soils and residual trees significantly.[55] RIL involves mapping desired trees and planning extraction strategies such as building roads efficiently and minimizing damaged woody biomass associated with logging. However, RIL is not widely implemented and faces several barriers in the tropics, including higher costs and illegal logging.[56]

Fuelwood extraction is done largely for subsistence.[57] Indigenous peoples and small farmers harvest fuelwood for cooking and heat. In some areas, large-scale harvesting is used to produce charcoal for subsistence and markets. In contrast to commercial logging, tree species used for fuel are largely irrelevant. Thus, there is less waste, but more complete clearing, which can lower the prospect for forest regeneration.

Governance

Government action (or inaction) can be an underlying cause of deforestation in several ways. For example, governments can fund and determine where roads are built; determine land rights and uses that affect forest clearing; influence enforcement of forest laws; and affect deforestation through tax policies, production subsidies, and other institutional choices (immigration and development policies).

- **Infrastructure development.** Government policies determine road building in forested areas. If roads are constructed without regard for environmental or development considerations, they can exacerbate deforestation. Policies can also temper deforestation stemming from new roads by restricting land uses in proximity to roads, or by creating forest reserves to manage logging and agriculture.

---

[54] Geist and Lambin, *What Drives Tropical Deforestation?* p. 25.

[55] See Dykstra, *Reduced Impact Logging.*

[56] See CRS Report RL33932, *Illegal Logging: Background and Issues,* by Pervaze A. Sheikh.

[57] Geist and Lambin, *What Drives Tropical Deforestation?* p. 29.

- **Land Rights/Tenure.** Secure title to land is important for implementing land use policies affecting deforestation. For example, farmers and ranchers with secure title to their land are more inclined to make investments on their land to practice permanent agriculture. Secure title can also provide an incentive for landowners to participate in programs that preserve forests. If landowners know they will receive long-term benefits from programs that pay for forest preservation on their land, they are more likely to join. Indeed, many advocate improved land tenure arrangements to assure that landowners can capture the benefits of forest management, including forest carbon sequestration.[58] In many tropical countries, the government holds title to the land—about two-thirds of all tropical forest area is government-owned.[59] Clear and secure private title to land can also reduce the practice of deforestation to acquire title. In several tropical countries, establishing ownership required "productive use of the land," most readily demonstrated by clearing the forest; this contributed significantly to tropical deforestation.[60] Forest clearing to obtain title also indirectly prevented land ownership for native peoples who use the forests for subsistence.

- **Enforcement.** The enforcement of laws directly or indirectly related to forested lands can affect deforestation. Weak enforcement can result in widespread degradation or deforestation and undermine the effectiveness of forest policies and laws. Weak enforcement might stem from lack of resources (money and trained personnel) or from bribes or collusion.[61] One consequence of weak law enforcement is illegal logging, which has become a multi-billion dollar enterprise. Illegal logging affects market prices for timber, depressing returns to legitimate landowners (including the government) and often leading to additional logging to generate the same level of revenues. In some places, illegal logging is rampant, accounting for as much as 80% to 90% of timber harvesting regionally.[62]

- **Institutional Factors.** Government policies related to market signals can be an underlying cause of tropical deforestation.[63] Agricultural subsidies are commonly cited, and include tax expenditures, import tariffs, fertilizer assistance, and other policies that alter market signals to encourage agricultural production, particularly for export.[64] A relationship between agricultural commodity prices and deforestation is strongly indicated by the roughly 50% decline in Amazonian deforestation in Brazil between 2008 and 2009,[65] when global beef prices fell

---

[58] M. Kanninen et al., *Do Trees Grow on Money? The Implications of Deforestation Research for Policies to Promote REDD*, infobrief, Center for International Forestry Research (CIFOR), Bogor, Indonesia, 2007.

[59] J. Hatcher and L. Bailey, *Tropical Forest Tenure Assessment: Trends, Challenges, and Opportunities*, Rights and Resources Initiative and International Tropical Timber Organization, Yaounde, Cameroon, May 2009, p. 16.

[60] D. C. Nepstad, C. M. Stickler, and O. T. Almeida, "Globalization of the Amazon Soy and Beef Industries: Opportunities for Conservation," *Conservation Biology*, vol. 20, no. 6 (2006), pp. 1595-1603.

[61] A. Pfaff et al., *Policy Impacts on Deforestation: Lessons from Past Experiences to Inform New Initiatives*, Nicholas Institute for Environmental Policy Solutions, NI PB 09-04, June 2009.

[62] See CRS Report RL33932, *Illegal Logging: Background and Issues*, by Pervaze A. Sheikh.

[63] Geist and Lambin, *What Drives Tropical Deforestation?* pp. 8-12, 37-40.

[64] Eliasch, *Climate Change: Financing Global Forests*, pp. 39-43; and Pfaff et al., *Policy Impacts on Deforestation*.

[65] National Institute for Space Research (Brazil), "INPE Confirms 7,0008 km2 of Deforestation by Shallow Cut in the Amazon State," November 12, 2009, http://www.inpe.br/ingles/news/news_dest90.php. The Institute's graph (http://www.inpe.br/noticias/arquivos/pdf/grafico1_prodes2009.pdf) shows 2009 deforested area at about half of the (continued...)

---

substantially.[66] Similarly, deforestation in Cameroon was shown to be strongly correlated with the price of cash crops, notably coffee and cacao.[67] Growth in demand for biofuels could induce landowners in tropical countries to clear forests to grow corn, soybeans, sugarcane, or palm oil to satisfy the world's demand for alternative fuels. Policies related to population and economic stability can affect deforestation, too. For example, Indonesia's transmigration policy—encouraging people to move from densely populated areas to rural areas by giving them land for agriculture—has indirectly led to deforestation.[68] Government policies that address population, economic development, and international trade can have significant effects on deforestation rates.

Regional Drivers of Deforestation

Drivers of deforestation in the tropics have varying characteristics and intensities that make them different among tropical regions. Differences are related to forest type, demand for agricultural products and biomass, access to forests, urbanization patterns, historic and cultural relations to forests and forestlands, and other factors.

Differing regional drivers of deforestation are important to understand, because they influence the framework of policies to reduce deforestation rates. For example, pending climate change legislation in the U.S. Congress would authorize funds and other incentives (e.g., carbon offsets) to assist developing countries to reduce deforestation. Some of these programs are expected to be operated at the national level in developing countries. Understanding the connection between their national policies and deforestation could make deforestation-reduction efforts more targeted and efficient. Other programs are contemplated at the project level. Understanding regional and local drivers of deforestation could facilitate implementation of these projects.

Latin America, including Amazonia

The tropical forests of Latin America stretch from Mexico to southern Brazil and Bolivia. The 588 million hectare Amazon River basin dominates discussions about tropical forests, since it accounts for about two-thirds of tropical forests in Latin America. About 62% of the Amazon basin is in Brazil, with tracts in Peru, Bolivia, Columbia, and other countries.[69] In 2006, the trees in the Amazon basin were estimated to contain one-fourth to one-half of all terrestrial carbon in vegetation.[70] Some have suggested that the Brazilian Amazon alone may contain as much as 40%

(...continued)

2008 area and the lowest level since monitoring began in 1988.

[66] About a 25% decline in the FAO Bovine Meat Index; see http://www.fao.org/es/ESC/en/15/138/highlight_583.html. Other factors might also have had an effect, such as increased law enforcement and greater restrictions on deforestation implemented by the Brazilian government during the same period.

[67] V. Bellassen et al., *Reducing Emissions from Deforestation and Degradation: What Contribution from Carbon Markets?*, Caisse des Dépôts, Climate Report: Issue n°14, Paris, France, September 2008.

[68] A. J. Whitten, "Indonesia's Transmigration Program and Its Role in the Loss of Tropical Rain Forests," *Conservation Biology*, vol. 1, no. 3 (October 1987), pp. 239-246.

[69] A. T. Wolf, L. de Silva, and K. Hatcher, *SOUTH AMERICA: International River Basin Register*, Oregon State University, Program in Water Conflict Management and Transformation, Corvallis, OR, August 2002, http://www.transboundarywaters.orst.edu/publications/register/tables/IRB_southamerica.html.

[70] B. S. Soares-Filho et al., "Modeling Conservation in the Amazon Basin," *Nature*, vol. 440 (May 2006), pp. 520-523.

of all remaining tropical rainforest.[71] An estimated average of 1.8 million hectares of Brazilian Amazon rainforest was lost annually between 1988 and 2008, about a third of global tropical deforestation.[72]

Anthropogenic drivers of deforestation in Latin America and especially the Amazon are generally centered on agricultural expansion, including both large-scale and small-scale agriculture. The primary driver of deforestation has been cattle ranching.[73] Some estimate that cattle ranching in the Amazon is associated with four-fifths of Amazon deforestation.[74] High global beef prices in the early 1990s led to an estimated 11% annual increase in the cattle herd from 1997 to 2003 and a surge in deforestation in 2002 to 2004.[75] Soybean production and more recently sugarcane and other crops for biofuels are also significant drivers of deforestation; in general, production of these biofuels is not directly causing deforestation, but rather is displacing cattle ranching, forcing cattle production further into forested areas.

The emergence of agriculture as a driver of deforestation in Brazil was initially tied to government incentives in the 1960s that promoted migration and development in the Amazon. Cheap land and equipment were used to motivate settlers to colonize the Amazon and stake claim to borderlands. Major highways connecting the south of Brazil to parts of the Amazon were also built to facilitate this effort. Roads and highways provided access to forests and enabled farmers and loggers to transport their goods to markets. Farmers practiced swidden agriculture because they found it cheaper to clear new land for planting crops than to maintain their original tracts. Their old land was generally abandoned or converted to pastureland for ranching. In recent years, market forces, particularly global beef prices, have been identified as a significant underlying driver of deforestation. Agricultural products coming out of the Amazon are entering the global market, and changes in domestic and global market conditions have been correlated to rates of deforestation.[76]

Deforestation in the Amazon is exacerbated by forest fires. Some contend that the frequency of drought is a prime determinant of how often forest fires occur and how extensive they become in the tropical forests of the Amazon.[77] Combined with land use activities and drought conditions, successive fires in the same area can prevent regeneration in the Amazon.[78]

Tropical Africa

The tropical forest ecosystems of Africa are centered around the Congo River basin, the second largest river basin in the world (after the Amazon), and the coast of the Gulf of Guinea. The

---

[71] J. A. Foley et al., "Amazonia Revealed: Forest Degradation and Loss of Ecosystem Goods and Services in the Amazon Basin," *Frontiers in Ecology and the Environment*, vol. 5, no. 1 (February 2007), pp. 25-32.

[72] A. S. L. Rodrigues et al., "Boom-and-Bust Development Patterns Across the Amazon Deforestation Frontier," *Science*, vol. 324, no. 5933 (June 2009), pp. 1435-1437.

[73] See Bellassen et al., *What Contribution from Carbon Markets?*

[74] D. C. Nepstad et al., "The End of Deforestation in the Brazilian Amazon," *Science*, vol. 326, no. 5958 (December 2009), pp. 1350-1351.

[75] Nepstad et al., "Globalization of the Amazon Soy and Beef Industries."

[76] T. J. Killeen, "A Perfect Storm in the Amazon Wilderness," *Advances in Applied Biodiversity Science*, vol. 7 (2007).

[77] Aragão et al., "Interactions Between Rainfall, Deforestation and Fires."

[78] Some studies report that repeated forest fires can cause the near-permanent conversion of forest to scrub or grassland. See, for example, Cochrane and Barber, "Climate Change, Human Land Use and Future Fires in the Amazon."

Congo basin spans six countries in Central Africa, covering about 369 million hectares.[79] It is the world's second-largest area of contiguous tropical rainforest, with more than 24 million people living in or around the forest and relying on it for agriculture, food, medicine, fuel, and construction materials.[80] The Congo Basin is surrounded by a band of dry tropical forests; sometimes classified as savannahs because of their relatively low tree cover. Deforestation rates in tropical Africa are considerably lower than in Asia and Latin America.[81] However, high population growth rates, coupled with the demand for agricultural land and woody biomass for fuel, could double the rate of deforestation in the next 20 years.[82]

Deforestation in both the wet and dry tropical forests of Africa is driven largely by small-scale agriculture, both shifting cultivation and permanent production of such crops as cassava, yams, plantains, and millet.[83] However, conversion to large-scale agriculture is an emerging threat to forests in the region, and accounts for about a third of deforestation in tropical Africa.[84] Cattle ranching is a much less significant driver of deforestation in Africa than in Latin America.[85]

Fuelwood, both for local use and for charcoal for urban use, has been a major cause of deforestation in Africa.[86] An estimated 90% of the continent's population uses fuelwood for cooking, and in sub-Saharan Africa, firewood and brush supply about 50% of all energy sources.[87] Some have noted that only 7.5% of rural households in sub-Saharan Africa have access to electricity, so demand for fuelwood is likely to continue to be a cause of deforestation in tropical Africa for the foreseeable future.[88]

Commercial logging is less significant in tropical Africa than elsewhere in the tropics.[89] Nonetheless, the effects of logging can still be significant by degrading forests and providing access for farmers. The annual clearing of dense forest is related to the rural population density near the forest, suggesting that proximity of the populations and their access to forests is a major cause of deforestation in the region.[90] Some use the access argument to contend that creating

---

[79] Wolf, de Silva, and Hatcher, *AFRICA: International River Basin register*, http://www.transboundarywaters.orst.edu/publications/register/tables/IRB_africa.html.

[80] U.S. Agency for International Development (USAID), *Congo Basin Forest Partnership*, http://www.usaid.gov/locations/sub-saharan_africa/initiatives/cbfp.html.

[81] E. F. Achard et al., "Determination of Deforestation Rates of the World's Humid Tropical Forests," *Science*, vol. 297, no. 5583 (August 2002), pp. 999-1002.

[82] Quanfa Zhang et al., "A GIS-Based Assessment on the Vulnerability and Future Extent of the Tropical Forests of the Congo Basin," *Environmental Monitoring and Assessment*, vol. 114 (2006), pp. 107-121.

[83] Y. Agyei, "Deforestation in Sub-Saharan Africa," *African Technology Forum* vol. 8, no. 1 (1998), http://web.mit.edu/africantech/www/articles/Deforestation.htm; Olander et al., *International Forest Carbon and the Climate Change Challenge*, p. 44; FAO, *State of the World's Forests, 2009*, pp. 5-6; and FAO data at http://faostat.fao.org/site/339/default.aspx.

[84] Project Catalyst, *Towards the Inclusion of Forest-Based Mitigation in a Global Climate Agreement*, Climate Works Foundation, 2009, p. 12; FAO, *State of the World's Forests 2009* (Rome, Italy: 2009), p. 5.

[85] Project Catalyst, *Towards the Inclusion of Forest-Based Mitigation in a Global Climate Agreement*. p. 12; and Geist and Lambin, *What Drives Tropical Deforestation?* p. 25.

[86] Geist and Lambin, *What Drives Tropical Deforestation?* p. 29.

[87] Agyei, "Deforestation in Sub-Saharan Africa."

[88] FAO, *State of the World's Forests, 2009*, p. 8.

[89] Geist and Lambin, *What Drives Tropical Deforestation?* p. 29.

[90] Quanfa Zhang et al., "Mapping Tropical Deforestation in Central Africa," *Environmental Monitoring and Assessment*, vol. 101, no. 1-3 (January 2005), pp. 69-83.

---

forest reserves in the middle of contiguous or unoccupied forests might be detrimental to conservation. They argue that roads built to access reserves for protection might also stimulate deforestation in areas surrounding the reserve, potentially resulting in greater forest loss than what the reserve would protect.[91]

Some underlying causes of deforestation in tropical Africa are related to governance. Examples include a low priority (which may translate into little funding) for forest conservation; poor enforcement of forest conservation laws; few incentives to conserve forests; treatment of forests as a commons areas; and lack of defined property rights.[92] These factors are magnified in Africa, since approximately 98% of tropical forests are managed by governments.[93] Poor law enforcement particularly affects forest reserves in tropical Africa.[94] Reserves are exposed to poachers, wood gatherers, and logging. Some contend that, with an increasing population and diminishing forest area, failure to enforce laws for protected areas and ensure land rights for local communities will impede efforts to check deforestation.[95]

Political instability, including wars and civil disturbances, in parts of Africa also weakens law enforcement, leading to greater rates of deforestation by the refugees and displaced persons as well as through commercial logging to finance the conflicts.[96] However, instability due to conflict may also restrict investment and infrastructure expansion, thus limiting deforestation in parts of the Congo basin.[97]

Southeast Asia

Southeast Asia includes the Indochinese Peninsula and the islands of Indonesia and the Philippines, as well as many other nearby islands. The entire region contains tropical rainforests. Approximately a third of the tropical forest area in the world is in Southeast Asia. Indonesia alone reportedly contains approximately 9% of the world's tropical forest area.[98]

Deforestation in Southeast Asia is largely driven by agriculture. In many areas, deforestation is similar to tropical Africa, being driven substantially by small-scale shifting and permanent agricultural plots.[99] Also, like Africa and unlike Latin America, cattle ranching is a minor factor in deforestation.[100] Another parallel to tropical Africa is that, except for Malaysia and Indonesia,

---

[91] D. B. Bray et al., "Tropical Deforestation, Community Forests, and Protected Areas in the Maya Forest," *Ecology and Society*, vol. 13, no. 2 (2008).

[92] L. P. Olander et al., *International Forest Carbon and the Climate Change Challenge: Issues and Options*, Nicholas Institute for Environmental Policy Solutions, June 2009, pp. 45-46.

[93] Hatcher and Bailey, *Tropical Forest Tenure Assessment*, p. 16.

[94] FAO, *State of the World's Forests, 2009*, pp. 4-5.

[95] Hatcher and Bailey, *Tropical Forest Tenure Assessment*.

[96] Scribd., *Ecosystem Services of the Congo Basin Forests*, http://www.scribd.com/doc/5501370/Ecosystem-Services-of-the-Congo-Basin-Forests.

[97] Eliasch, *Climate Change: Financing Global Forests*, p. 36.

[98] M. C. Hansen et al., "Humid Tropical Forest Clearing from 2000-2005 Quantified by Using Multitemporal and Multiresolution Remotely Sensed Data," *Proceedings of the National Academy of Sciences*, vol. 105, no. 27 (2008), pp. 9439-9444.

[99] Geist and Lambin, *What Drives Tropical Deforestation?* p. 25; and FAO, *State of the World's Forests, 2009*, p.15.

[100] Geist and Lambin, *What Drives Tropical Deforestation?* p. 25.

commercial logging is a relatively modest cause of deforestation.[101] However, in Papua New Guinea, logging is a major driver in lowland forests, with subsistence agriculture causing deforestation throughout the rest of the country.[102]

Deforestation in Indonesia and Malaysia differ from most of the rest of Southeast Asia. Forests are extensive, and deforestation rates have been relatively high. Indonesia accounts for nearly 13% of global tropical deforestation.[103] Extensive commercial logging has been a major contributor to high rates of deforestation in both Indonesia and Malaysia.[104] However, in recent years, large-scale commercial agriculture is the predominant cause of deforestation.[105] Oil palm is the major commercial crop in both countries, and extensive areas have been cleared for oil palm plantations.[106] Deforestation also occurs for rubber plantations, which have been planted extensively in Thailand as well as in Malaysia.[107]

Some underlying drivers of deforestation in Southeast Asia include government policies and market forces. In Indonesia, for example, federal policies encouraged resettlement from urban centers into forested areas in the 1980s.[108] More recently, deforestation in Indonesia has been more enterprise-driven, primarily caused by conversion to agriculture (e.g., oil palm plantations).[109] Some also argue that industrial expansion and development in China is also an underlying force driving deforestation in Southeast Asia.[110] Increasing demand for tropical hardwoods, palm oil, and rubber for consumption in China are thought to be driving deforestation in Indonesia and Malaysia. An expanding Chinese economy could sustain this demand and lead to greater deforestation in Southeast Asia, despite government policies and incentives to reduce deforestation.[111]

Climate Consequences of Tropical Deforestation

Most scientists agree that, in the past two decades, tropical deforestation has been responsible for the largest share of CO_2 released to the atmosphere from land use changes.[112] At current rates of

---

[101] Geist and Lambin, *What Drives Tropical Deforestation?*; and Bellassen et al., *Reducing Emissions from Deforestation and Degradation.*

[102] See section on Papua New Guinea at http://rainforests.mongabay.com/20png.htm.

[103] Hansen et al., "Humid Tropical Forest Clearing."

[104] Geist and Lambin, *What Drives Tropical Deforestation?* p. 29; and Olander et al., *International Forest Carbon and the Climate Change Challenge*, p. 44.

[105] Olander et al., *International Forest Carbon and the Climate Change Challenge*, p. 44.

[106] T. K. Rudel, "Changing Agents of Deforestation: From State-Initiated to Enterprise Driven Processed, 1970-2000," *Land Use Policy*, vol. 24, no. 1 (2007), pp. 35-41; FAO, *State of the World's Forests, 2009*, p. 15; and FAO data at http://faostat.fao.org/site/339/default.aspx.

[107] S.A. Abdullah and N. Nakagoshi, "Changes in Agricultural Landscape Pattern and Its Spatial Relationship with Forestland in the State of Selangor, Peninsular Malaysia," *Landscape and Urban Planning*, vol. 87, no. 2 (August 2008), pp. 147-155; and FAO data at http://faostat.fao.org/site/339/default.aspx.

[108] The government-initiated Transmigration Project led to the migration of approximately 2 million people from Java to forested areas in Irian Jaya, Sumatra, and Kalimantan. See Whitten, "Indonesia's Transmigration Program"; and Williams, *Deforesting the Earth*, p. 459.

[109] Rudel, "Changing Agents of Deforestation: From State-Initiated to Enterprise Driven Processes."

[110] S. Kemper, "Forest Destruction's Prime Suspect," *Environment Yale*, Spring 2008, pp. 4-11.

[111] Kemper, "Forest Destruction's Prime Suspect."

[112] IPCC 2007 WG-I Report.

deforestation, clearing tropical forests could release an additional 87 to 130 GtC of CO_2 to the atmosphere by 2100.[113]

Tropical deforestation can quickly deplete the moderate levels of soil carbon. In rainforests, heavy precipitation leaches carbon (and other minerals) into the surface waters and the groundwater. This is, in part, why shifting agriculture is a common practice in tropical areas—soil nutrients (including carbon) are quickly depleted, reducing vegetative growth and requiring farmers to find new land. This carbon depletion also hinders regrowth of the forest after croplands are abandoned.

The soil carbon impacts of tropical deforestation are particularly important for Indonesia, because of its relatively extensive peatlands.[114] Peat soils have a very high carbon content, typically because they oxidize very slowly as a result of often being flooded. Deforestation, and drainage for commercial oil palm plantations, releases large amounts of carbon in a relatively short period.[115] Thus, deforesting and draining peat forests have particularly significant impacts on global carbon sequestration.

Commercial logging in the tropics also affects climate through the relatively large release of carbon, compared to logging in temperate and boreal forests. The emphasis of commercial wood production on only a relatively few tree species in tropical forests often results in substantial waste—harvests may take as little as 10% of the wood volume, and many non-target trees are killed or damaged.[116] Furthermore, commercial logging usually includes roads for removing the timber, the first step in general access for agricultural expansion and other developmental (forest-clearing) activities.

Fire can exacerbate the climate impacts of tropical deforestation. The natural burning regime in tropical rainforests differs from that of boreal and temperate forests. Natural fires are relatively rare in moist tropical forests, with natural fire cycles measured in hundreds or even thousands of years.[117] However, fire is commonly used (often following commercial logging) to clear lands for agriculture—crops or pastures. This releases the carbon from the vegetation that is cut down. Much of this is released to the atmosphere, but some temporarily adds nutrients to the soils, increasing plant growth for a few years.[118] However, land-clearing fires often escape and burn surrounding forests. Fires in the tropics are harmful in three ways: (1) they release substantial quantities of CO_2 to the atmosphere; (2) they generate substantial volumes of smoke, causing "brown clouds" and regional health problems;[119] and (3) they create a positive feedback loop by

[113] R. A. Houghton, "Chapter 1: Tropical Deforestation as a Source of Greenhouse Gas Emissions," in *Tropical Deforestation and Climate Change*, ed. P. Moutinho and S. Schwartzman (Washington: Amazon Institute for the Environment (IPAM) and Environmental Defense Fund, 2005), pp. 13-23.

[114] See Y. Uryu et al., *Deforestation, Forest Degradation, Biodiversity Loss and CO_2 Emissions in Riau, Sumatra, Indonesia*, World Wildlife Fund, WWF Indonesia Technical Report, Jakarta, Indonesia, February 7, 2008.

[115] P. A. Minang et al., eds, "The Opportunity Costs of Avoiding Emissions from Deforestation," ASB PolicyBrief No. 10, ASB Partnership for the Tropical Forest Margins, Nairobi, Kenya.

[116] Tropical Forest Foundation, "Reduced Impact Logging."

[117] M. A. Cochrane, "In the Line of Fire: Understanding the Impacts of Tropical Forest Fires," *Environment*, vol. 43, no. 8 (October 1, 2001), p. 28.

[118] L. Tacconi, P. F. Moore, and D. Kaimowitz, "Fires in Tropical Forests—What Is Really the Problem? Lessons from Indonesia," *Mitigation and Adaptation Strategies for Global Change*, vol. 12, no. 1 (January 2007), pp. 55-66.

[119] Ramanathan et al., "Warming Trends in Asia Amplified by Brown Cloud Solar Absorption."

opening and drying the adjoining forests.[120] This perpetuates a cycle of burning, since post-fire habitats (other than crop and pasture lands) are usually dominated by flammable grasses and vines that make the area susceptible to more destructive and more extensive fires.[121]

Finally, tropical deforestation has also been shown to be linked to decreased evapotranspiration, which lessens atmospheric moisture and precipitation levels.[122] As noted above, this could reduce precipitation and increase surface temperatures regionally. Warming in tropical regions could increase the susceptibility of tropical forests to fires and increase tree mortality due to drought, as discussed earlier. In addition, reduced precipitation might reduce agricultural productivity, leading to increased deforestation simply to maintain agricultural output levels.

Reducing Deforestation

The drivers of deforestation suggest various approaches to reducing deforestation: adjusting markets and assisting tropical countries with infrastructure and governance. For a description of U.S. programs that address deforestation, see the **Appendix**. In addition, the net effect of deforestation might, in some circumstances, be offset by afforestation or reforestation—planting trees on the cleared sites.

Tree Planting to Offset Deforestation?

Two related forestry practices are sometimes proposed to mitigate deforestation: afforestation (establishing trees on sites that have *long* been cleared of forests, such as crop, pasture, and brush lands); and reforestation (establishing tree stands on areas *recently* cleared or partially cleared of forest through timber harvesting or natural causes).[123] Afforestation of crop or pasture land has been one focus of attention for carbon sequestration by domestic stakeholders.[124]

Some have suggested that the additional carbon sequestration from afforestation and reforestation could offset the carbon release from deforestation. They assert that harvesting "over-mature" forests sequesters additional carbon, because (1) very old forests sequester little additional carbon (the amount stored is large, but the annual addition is small or even negative); (2) wood products made from the timber continue to store carbon for decades; and (3) newly established stands grow vigorously, sequestering large amounts of carbon.[125]

---

[120] D. C. Nepstad, *The Amazon's Vicious Cycles: Drought and Fire in the Greenhouse*, World Wide Fund for Nature (WWF), 2007.

[121] Cochrane, "Understanding the Impacts of Tropical Forest Fires."

[122] Bala et al., "Combined Climate and Carbon-Cycle Effects of Large-Scale Deforestation."

[123] There are no definitive periods for "long cleared" and "recently cleared," but establishing new tree stands on sites cleared within the past decade is usually considered reforestation. This distinction has little relevance for on-site practices, since the activities for afforestation and reforestation do not differ. However, the distinction may be significant under the various protocols for reducing GHG emissions.

[124] See CRS Report R41086, *Potential Implications of a Carbon Offset Program to Farmers and Landowners*, by Renée Johnson et al.

[125] J. Perez-Garcia, C. D. Oliver, and B. Lippke, "How Forests Can Help Reduce Carbon Dioxide Emissions to the Atmosphere," U.S. Congress, House Resources, Forests and Forest Health, *Hearing on H.Con.Res 151*, 105th Cong., 1st sess., September 18, 1997, Serial No. 105-61 (Washington: GPO, 1998), pp. 46-68.

Others dispute these claims, asserting that harvesting old-growth forests (commonly described by foresters as "over-mature") results in a net release of carbon.[126] Researchers have determined that carbon continues to accumulate in old-growth forests for centuries, long after the traditional definition of over-mature.[127] Other research has found that some old-growth forests continue to accumulate carbon in the soil.[128] Finally, the limited research evidence has shown that *intact* (uncut) natural forests store much greater volumes of carbon than do mature plantations—as much as three times as much carbon.[129]

Both of these conclusions may be valid in certain circumstances, depending on factors such as which products are manufactured, how those products are used, how much carbon is left on the site, and what happens to it. There are, of course, other considerations (e.g., the impacts on ecosystem services and on local economies) associated with discussions of harvesting old-growth forests. It should be recognized that these arguments have been made with respect to old-growth forests in temperate and boreal regions. Timber harvesting (and other forest clearing) in these regions likely contribute relatively modest carbon emissions to the atmosphere, because fewer species dominate temperate and boreal forests, and reforestation commonly follows the harvests. In addition, relatively few old-growth forests remain in these regions.

The situation differs in tropical forests. Because of the wide diversity of plant species and the emphasis of commercial wood production on relatively few tree species, timber harvesting in tropical forests often results in substantial waste, with a smaller portion of the wood volume removed and substantial damage to many non-target trees.

Furthermore, as noted above, commercial logging often opens the forests for agriculture, and the soil depletion and burning associated with agriculture may prevent effective forest recovery in the tropics. Naturally regrown (second-growth) forests in the tropics have been shown to contain less biological diversity and less total biomass (carbon) than intact forests, and forest plantations in the tropics have far less diversity and biomass than second-growth forests.[130] Thus, under many circumstances, deforestation in tropical forests emits substantial quantities of carbon that cannot be adequately compensated by reforestation except in the very long term (several decades to centuries).

Market Solutions

There are basically three market approaches to reducing deforestation: specific markets for forest carbon; general markets for ecosystem services and non-timber forest products; and certified sustainable forestry. These approaches are not entirely independent or mutually exclusive choices;

---

[126] M. E. Harmon, W. K. Ferrell, and J. F. Franklin, "Effects on Carbon Storage of Conversion of Old-Growth Forests to Young Forests," *Science*, vol. 247 (February 9, 1990), pp. 699-702; and P. M. Vitousek, "Can Planted Forests Counteract Increasing Atmospheric Carbon Dioxide?" *Journal of Environmental Quality*, v. 20 (Apr.-June 1991), pp. 348-354.

[127] S. Luyssaert et al., "Old-Growth Forests as Global Carbon Sinks," *Nature*, vol. 455 (September 11, 2008).

[128] Guoyi Zhou et al., "Old-Growth Forests Can Accumulate Carbon in Soils," *Science*, v. 314 (Dec. 1, 2006), p. 1417.

[129] B. G. Mackey et al., *Green Carbon: The Role of Natural Forests in Carbon Storage, Part 1, A Green Carbon Account of Australia's South-Eastern Eucalypt Forests, and Policy Implications*, The Fenner School of Environment & Society, The Australian National University, Canberra, Australia, 2008.

[130] E. Stokstad, "Forests in Flux: A Second Change for Rainforest Biodiversity," *Science*, vol. 320 (June 13, 2008), pp. 1436-1438.

---

for example, the carbon benefits from certified sustainable forestry practices might be salable in forest carbon markets.

Forest Carbon Markets[131]

Carbon markets have formed to encourage voluntary efforts to reduce GHG emissions as well as to fulfill mandatory or regulatory GHG emission reductions. Forestry activities, including reduced deforestation, might generate carbon credits or offsets—reductions in GHG emissions or increases in carbon sequestration that regulated entities (or volunteers) can purchase to offset the GHGs for which they are responsible.

Various forestry practices have been considered for carbon credits.[132] Each practice has carbon benefits, but there are also concerns and limitations in allowing these offsets for mandatory GHG reduction programs. Concerns include:

- Would the forest activity add to the effort to reduce CO_2 emissions, or would the activity have occurred anyway under existing laws and practices? (This issue is referred to as "additionality.")

- How much carbon would have been released if no action (e.g., to prevent deforestation) had occurred? (This issue is referred to as defining or determining "the baseline.")

- Can the carbon sequestration benefits of a project be verified? Verifying project benefits for every project and monitoring benefits over time can be expensive, and can be further restricted because of lack of access to forests. (This issue is referred to as "measurement, monitoring, and verification.")

- If a GHG reduction project is implemented, will the GHG-emitting practice or activity (e.g., deforestation) shift to a different location or country with no GHG reduction policies? (This issue is referred to as "leakage.")

- Will the project result in permanent reductions in GHG concentrations, or will the effects be temporary? (This issue is referred to as "permanence.")

Avoided tropical deforestation offers a mixture of benefits and difficulties associated with implementation. Areas already legally protected (e.g., national parks and reserves) would generally not qualify, because their carbon sequestration would not be additional. However, they might be susceptible to leakage from poachers and squatters that practice illegal deforestation. Also, natural disasters (wildfires, hurricanes, etc.) can effectively destroy forests, releasing their carbon, and accounting for such releases complicates efforts to assure the "permanence" of forest carbon.

---

[131] See CRS Report RL34560, *Forest Carbon Markets: Potential and Drawbacks*, by Ross W. Gorte and Jonathan L. Ramseur.

[132] See CRS Report RS22964, *Measuring and Monitoring Carbon in the Agricultural and Forestry Sectors*, by Ross W. Gorte and Renée Johnson.

Markets for Ecosystem Services and Non-Timber Forest Products

Ecosystem or environmental services encompass a wide variety of benefits, including carbon storage. Forests and other undeveloped lands provide a host of environmental services, such as climate regulation, soil retention, waste remediation, and clean water. Landowners generally are not compensated for these services. Some have sought ways to provide such compensation as an incentive to landowners to keep their lands forested. Forest carbon markets are special ecosystem services markets that could compensate landowners for the carbon storage services their forests provide.

Payments for ecosystem services (PES) is an approach where beneficiaries of the services are identified and then charged to pay landowners to maintain their forests. A PES program was established in Costa Rica in the 1990s. About half of the country was deforested from the 1950s through the 1980s. Tax-funded subsidies to prevent deforestation were initially successful, but were not politically and financially sustainable. The subsidy program was then replaced with a legal framework banning deforestation and requiring users of forest services to pay to restore and protect Costa Rican forests. Initially, user fees were established for wood fuel, then expanded for water supplies. Public support for the PES program was generated by clearly establishing the linkage between the services and the users, and by setting the fees to not only compensate the landowners for not deforesting their lands but also cover administrative costs (e.g., inventory and monitoring costs). While the program has reduced deforestation, it has not been sufficient to eliminate forest losses.

A long-standing U.S. example is buying duck stamps (essentially a federal duck-hunting permit) in order to hunt ducks; money from the stamps is used to conserve duck habitat, which makes more hunting possible.[133] The concept of creating ecosystem services markets is being pursued by the USDA under authority provided in the 2008 farm bill (P.L. 110-246, the Food, Conservation, and Energy Act of 2008).[134] To the extent that ecosystem or environmental services markets develop more broadly, with or without federal support, they will likely encompass forest (and perhaps also soil) carbon sequestration among the services for which landowners are compensated.

A related market is for harvesting non-timber forest products without cutting down the trees. Such products include exotic nuts and berries, wild mushrooms (e.g., morels), natural rubber, floral greenery, and more. Markets for non-traditional, non-timber products harvested from forests have been growing around the world. While non-timber products will probably never supplant commercial timber values, the products can often be harvested with minimal impact on the forest and with virtually no carbon release. Encouraging additional growth in the markets for non-timber forest products can provide landowners with incentives to keep their forests intact, thereby contributing to long-term carbon sequestration.

---

[133] The program is not strictly comparable to PES programs, since the landowners rarely receive payments for providing duck habitat; rather, the duck stamps provide funding for state agencies to provide habitat and other benefits for duck hunters.

[134] See CRS Report RL34042, *Provisions Supporting Ecosystem Services Markets in U.S. Farm Bill Legislation*, by Renée Johnson.

Certified Sustainable Forestry

Certified sustainable forestry is a market approach to reduce carbon release from net deforestation through sustainable forest management. Several certification systems exist, with significant differences in the parameters that must be met.[135] Certification can be based on management practices that allow for sustainable logging to maximize carbon stores and minimize collateral damage to neighboring trees. Landowners could benefit from consumer willingness to pay higher prices for wood products grown and harvested using sustainable practices. Most systems require chain-of-custody reporting to assure that wood products claiming to be from sustainable forests actually come from certified forest lands. While many forestland owners believe that the costs of becoming and remaining certified are less than the benefits of higher prices and consumer awareness, it is not yet clear that the price differential for certified wood products will be sufficient in the long run to maintain the certification systems.

Governance Issues

Presuming that a developing country wants to reduce deforestation—for domestic benefits and/or to participate in forest carbon markets—various governance issues might need to be addressed. Countries have different needs and capacities, and thus might need to address a few or many governance issues, and the effort required might be modest or substantial. As discussed above, as well as extensively in the literature, the array of governance issues that could be addressed to reduce deforestation includes:

- **Agricultural subsidies and policies.** Various government programs and policies keep input prices artificially low, provide tax incentives for cash crops, and otherwise alter market signals. Eliminating or reducing programs that encourage agricultural production at the expense of forests could reduce deforestation.[136]

- **Roads and infrastructure.** Roads and other public services (e.g., water and power) are critical for human expansion into forests. Roads provide access, which can contribute to deforestation. Infrastructure planning and development can reduce the level of deforestation by concentrating development (agriculture, forestry, and other activities) in already accessible areas.[137]

- **Land tenure and property rights.** Many studies have identified ill-defined tenure and property rights as a cause of deforestation, and have proposed explicit, clearly defined private land ownership as a means of reducing deforestation by giving individuals an ownership interest in the condition of their forestland.[138] This could reduce illegal logging and clearing by squatters (especially for swidden agriculture).

---

[135] B. Cashore et al., eds., *Confronting Sustainability: Forest Certification in Developing and Transitioning Countries*, Report No. 8 (New Haven, CT: Yale School of Forestry and Environmental Studies, 2006).

[136] Bellassen et al., *Reducing Emissions from Deforestation and Degradation*, and Pfaff et al., *Policy Impacts on Deforestation*.

[137] Pfaff et al., *Policy Impacts on Deforestation*.

[138] See, for example, L. Cotula and J. Mayers, *Tenure in REDD: Start-Point or Afterthought?*, International Institute for Environment and Development, Natural Resource Issues No. 15, London, UK, 2009, and Pfaff et al., *Policy Impacts on Deforestation*.

- **Forest-dependent communities and indigenous peoples.** Most observers recognize that, to reduce deforestation, the people who derive their living from forest resources must be involved.[139] While this may include land tenure and property rights, it also goes beyond, by responding to the interests and concerns of people affected by decisions about deforestation.

- **Enforcement.** Enforcement of deforestation policies is critical to effect change. This includes enforcing land tenure and property rights, protecting forest communities and indigenous peoples from squatters and other interlopers, halting construction of unplanned roads, preventing illegal logging, and more. Increased enforcement and oversight could also reduce corruption by increasing the transparency and visibility of forestland transactions.[140]

The capacity of tropical countries to address these governance issues varies widely. Some countries have already taken many steps; others are limited by poverty, population growth, and other factors. Financial and technical assistance from developed nations can help developing countries to establish and expand their capacities to address these governance issues. (See the **Appendix** for a description of existing U.S. programs that provide financial and technical forestry assistance.) In addition, developed countries can encourage improved governance in developing tropical countries through other methods, such as conditional loans (e.g., loans requiring actions by the borrower), debt relief (e.g., exchanging foreign debt for conservation actions), and demand management (e.g., banning illegally harvested timber).[141]

Forest and Deforestation Data Issues

Various sources report data on forest area and deforestation. However, the data differ, sometimes substantially. Sources have noted the discrepancies among reported data.[142] Why are the data discrepancies so substantial, even in relatively developed areas (e.g., the United States), where one might expect relatively high-quality data? There are two principal reasons: the classification of forest lands, and the measurement and reporting systems used.

Table 2 presents data from two sources that cover most of the world's forests. The U.N. Food and Agriculture Organization (FAO) has been assessing global forests for decades; its most recent report is the *Global Forest Resources Assessment 2005.*[143] The World Resources Institute (WRI), in cooperation with the U.N. Development Programme, the U.N. Environment Programme, and the World Bank, has also published data on global forests, in *World Resources, 2002-2004: Decisions for the Earth: Balance, Voice, and Power.*[144]

---

[139] See, for example, A. Agrawal, A. Chhatre, and R. Hardin, "Changing Governance of the World's Forests," *Science*, vol. 320 (June 2008), pp. 1460-1462; and Olander et al., *International Forest Carbon and the Climate Change Challenge*, pp. 25-27.

[140] Pfaff et al., *Policy Impacts on Deforestation.*

[141] Pfaff et al., *Policy Impacts on Deforestation.*

[142] See P. E. Waggoner, *Forest Inventories: Discrepancies and Uncertainties*, Resources For the Future, RFF DP 09-29, Washington, DC, August 2009.

[143] FAO Forestry Paper 147 (Rome, Italy, 2006).

[144] World Resources Institute, *World Resources, 2002-2004: Decisions for the Earth: Balance, Voice, and Power*, Washington. July 2003, pp. 270-271, http://www.wri.org/publication/world-resources-2002-2004-decisions-earth-balance-voice-and-power.

Table 2. Forest Area by Country or Region, by Predominant Forest Biome

(Acreage in million hectares; annual percent change from 1990 to 2005)

Countries/Regions	WRI 2000 Area	FAO: 2005 Area	FAO % Change, 1990-2005
Russia	851.4	808.8	- 0.02%
Canada	244.6	310.1	0.0
Sweden	27.1	27.5	+ 0.6%
Finland	21.9	22.5	+ 1.4%
Norway	8.9	9.4	+ 2.8%
Boreal Countries total	**1,153.9**	**1,178.3**	**+ 0.05%**
United States	226.0	303.1	+ 1.5%
China	163.5	197.3	+ 25.5%
Australia	154.5	163.7	- 2.5%
Europe[a]	126.0	133.2	+ 9.5%
Western and central Asia[b]	57.1	52.2	- 3.4%
Argentina	34.6	33.0	- 6.4%
Japan	24.1	24.9	- 0.3%
Other temperate forest countries[c]	61.2	61.8	-1.8%
Temperate Countries total	**847.0**	**969.7**	**+ 5.1%**
Brazil	543.9	477.7	- 8.1%
Democratic Republic of Congo	135.2	133.6	- 4.9%
Indonesia	105.0	88.5	- 24.1%
Central Africa[d]	92.4	89.1	- 4.6%
Southeast Asia[e]	71.6	83.2	- 7.4%
Tropical south Africa[f]	74.4	73.4	- 11.6%
West Africa[g]	72.4	74.6	- 16.3%
Other tropical South America[h]	72.8	67.3	- 7.8%
India	64.1	68.7	+ 5.9%
Peru	62.2	68.7	- 2.0%
Sudan	61.6	67.5	- 11.6%
Angola	69.8	59.1	- 3.1%
Mexico	55.2	64.2	- 6.9%
Bolivia	53.1	58.7	- 6.5%
Columbia	49.6	60.7	- 1.2%
Other east Africa[i]	48.1	54.9	- 11.1%
Venezuela	49.5	47.7	- 8.3%
Tanzania	38.8	35.3	- 14.9%
Zambia	31.2	42.5	- 13.6%
Burma[j]	34.4	32.2	- 17.8%

Countries/Regions	WRI 2000 Area	FAO: 2005 Area	FAO % Change, 1990-2005
Papua New Guinea	30.6	29.4	- 6.6%
Central America & Caribbean	23.5	28.4	- 14.0%
Other tropical countries[k]	13.7	11.2	- 7.5%
Tropical Countries total	**1,853.2**	**1,802.1**	**- 8.8%**

Sources:

World Resources Institute (WRI), *World Resources, 2002-2004: Decisions for the Earth: Balance, Voice, and Power*, with the U.N. Development Programme, the U.N. Environment Programme, and the World Bank, Washington, DC, July 2003, pp. 270-271, http://www.wri.org/publication/world-resources-2002-2004-decisions-earth-balance-voice-and-power.

Food and Agriculture Organization of the United Nations (FAO), *Global Forest Resources Assessment 2005*, FAO Forestry Paper 147, Rome, Italy, 2006, ftp://ftp.fao.org/docrep/fao/008/A0400E/A0400E00.pdf.

Notes: The countries may also contain forests of other biomes, but are reported in the forest biome in which they have the predominance of acreage. For example, the United States contains 51.3 million acres of boreal forest in Alaska (W. B. Smith et al., *Forest Resources of the United States, 2002*, USDA Forest Service, Northern Research Station, Gen. Tech. Rept. NC-241, St. Paul, MN, 2004, Table 1, p. 30), but is reported with temperate forests because five-sixths of U.S. forests are temperate forests. Countries with fewer than 1 million hectares (FAO data) are not identified with their region.

a. Excludes European countries with boreal forests: Finland, Norway, Russia, and Sweden.

b. Includes Afghanistan, Iran, Mongolia, Pakistan, Turkey, Turkmenistan and the other Asian former Soviet republics, and most of the Middle East.

c. Includes Bhutan, Chile, Democratic Republic of Korea, Nepal, New Zealand, north Africa (Egypt to Morocco and Western Sahara), Republic of Korea, and Republic of South Africa.

d. Includes Cameroon, Central African Republic, Equatorial Guinea, Gabon, and Republic of Congo.

e. Includes Cambodia, Laos, Malaysia, Philippines, Thailand, and Vietnam.

f. Includes countries south of Angola, Democratic Republic of Congo, and Tanzania.

g. Includes Mauritania, Mali, Niger, Nigeria, and countries to the south and west.

h. Includes Columbia, Ecuador, French Guiana, Guyana, Paraguay, and Suriname

i. Includes Chad, Eritrea, Ethiopia, Kenya, Somalia, and Uganda.

j. In FAO documents, this country is called Myanmar.

k. Includes Saudi Arabia, Solomon Islands and other Oceania countries, and Sri Lanka.

Table 2 shows that, although generally similar, the data do not match. For example, for the two most forested countries in the world, Russia and Brazil, WRI reported more forest area (5% and 14% more, respectively) than FAO reported. In contrast, FAO reported substantially more forest area in the United States (34% more) and Canada (27% more) than WRI reported. Similarly, data on deforestation amounts and rates differ widely. For example, the FAO data show Brazil accounting for 27% of tropical forests and 24% of tropical deforestation. Other data, limited to humid tropical forests (and thus excluding many African and Brazilian tropical forests), also show Brazil accounting for 27% of tropical forests, but 48% of tropical deforestation.[145] In addition, the FAO and WRI data by biome differ from the forest biome area data from the IPCC, shown in **Table 1**.

---

[145] Hansen et al., "Humid Tropical Forest Clearing."

The accuracy of the FAO data, especially deforestation rates, has particularly been questioned. One observer has noted "inconsistencies" in "three successively corrected declining trends" in FAO reports on forested areas.[146] This researcher argues that measurement errors, as well as changes in the statistical design and new data sources, raise serious questions about the reliability of the reported trends. FAO has acknowledged changes in reported acreages because of changes in standards for measuring forests.[147]

Classification of Forest Land

Since climate impacts vary by forest biome, classifying forests by biome is useful for assessing possible effects. One difficulty with forest classification is determining in which biome a forest belongs. While some of this may seem apparent—tropical forests can be defined as those between the Tropic of Cancer and the Tropic of Capricorn—the reality is that forests fall across a gradient of characteristics. For example, the subtropical forests of south Florida exhibit many traits in common with the tropical forests that, technically, occur a few degrees closer to the equator. Similarly, the distinction between temperate and boreal forests, while apparent through "classical" types, can be imprecise, with the typically temperate northern hardwood (maple-beech-birch) ecosystem mixing with the traditional boreal spruce-fir (mixed with birch) in the northern Lake States and New England and in southern Canada.

This biome classification is further complicated by forest area data being reported by country. Many countries can readily be assorted into particular biomes, such as the tropical forests of Brazil, the Democratic Republic of Congo, and Indonesia. However, many other countries straddle the imperfect boundaries between biomes. Russia, for example contains more than 800 million hectares of forest; most are the vast boreal forests of Siberia, but many are temperate forests in Europe. The more than 200 million hectares of forest in the United States is largely temperate, but includes extensive boreal forests in Alaska (perhaps a quarter of the total) as well as some tropical forests in Hawaii and Puerto Rico. Australia, with 30 million hectares, is similarly largely temperate, but the northern third or so is tropical forests. Thus, aggregating forest lands by biome is imprecise at best.

A more significant, but perhaps less obvious, classification problem is determining what constitutes a forest. Numerous definitions are used by different organizations in various places for a variety of purposes. The relevant measures include:

- **Trees.** The plants must be considered trees for the area to be considered a forest. There is no precise, botanical definition of a tree. Trees are perennial plants that typically grow with a single woody stem. Some sources specify minimum heights and/or diameters at maturity. In contrast, bushes and shrubs commonly have multiple woody stems. However, these distinctions are imprecise, at best; for example, aspen trees in a stand are commonly clones, with dozens of stems from a single rootstock, while bamboo is biologically a grass.

---

[146] A. Grainger, "Difficulties in Tracking the Long-Term Global Trend in Tropical Forest Area," *Proceedings of the National Academy of Sciences*, vol. 105, no. 2 (January 15, 2008), pp. 818-823.

[147] FAO Forestry Department, *FRA 2000—Comparison of Forest Area Change Estimates Derived from FRA 1990 and FRA 2000*, Working Paper 59, Rome, Italy, 2001, pp. 44-46, ftp://ftp.fao.org/docrep/fao/006/ad068e/AD068E.pdf.

- **Tree height.** The plants must be tall enough to be considered trees, at least at maturity. This might seem obvious, but in some settings (e.g., near timberline or at desert edges), trees can be quite short (1-2 meters tall). The FAO *1990 Global Forest Resources Assessment* defined forests as having trees at least 7 meters (23 feet) tall, while the *2000 Global Forest Resources Assessment* required trees at least 5 meters (16 feet) tall.[148]

- **Canopy closure.** A portion of the area must be covered by trees. While this could be measured by number of trees per hectare, a minimum percentage of the area covered by tree canopy is more common. The *1990 Global Forest Resources Assessment* defined forestlands as having at least 20% canopy cover (i.e., at least 20% of the area covered by tree crowns), while the *2000 Global Forest Resources Assessment* used 10% canopy cover.[149] In assessing forest habitats for the northern spotted owl (an admittedly narrow definition of "forest"), one group recommended 40% canopy closure in trees of at least 11 inches in diameter.[150] This also leads to questions of whether canopy cover is, and should be, a distinction between forests and woodlands (areas with some trees typically growing in arid or semi-arid grasslands).

- **Growth rate.** The site must be capable of growing trees (on human time scales). The U.S. Forest Service has long used a forest standard of lands capable of growing at least 20 cubic feet of commercially usable wood per acre per year (nearly 50 cubic feet of usable wood per hectare per year).[151]

One source reported that more than 650 definitions of forest were used in compiling the *2000 Global Forest Resources Assessment.*[152] While the definitions were similar in many ways, their application could alter forest area in a country by as much as 10%.

The classification of forest lands is further complicated by plantations and orchards. Apples, peaches, rubber, and coffee can be classified as perennial crops—agricultural lands, rather than forests. But what about plantations for lumber—pine, mahogany, teak, and the like? Many such plantations could be classified as forests, especially if the tree species are native to the area and relatively few efforts are required to control undesirable competing vegetation. Pulp and woody biomass energy plantations are more problematic—sometimes native species are used, but the plants might not be grown to tree sizes.

Measurement and Reporting Systems

The other primary cause of discrepancies in reports on forest acreage lies in the ways forests are measured and reported. Forest area is typically determined from maps, generated by on-site census or surveys or from aerial or satellite images.[153] For all but a few forest areas, censuses are

[148] FAO, *Comparison of Forest Area Change Estimates.*

[149] FAO, *Comparison of Forest Area Change Estimates.*

[150] J. W. Thomas et al., *A Conservation Strategy for the Northern Spotted Owl*, Interagency Scientific Committee to Address the Conservation of the Northern Spotted Owl, Portland, OR, May 1990.

[151] W. B. Smith et al., *Forest Resources of the United States, 2002*, USDA Forest Service, Northern Research Station, Gen. Tech. Rept. NC-241, St. Paul, MN, 2004.

[152] A. S. Mather, "Assessing the World's Forests," *Global Environmental Change*, vol. 15 (2005), pp. 267-280.

[153] See CRS Report RS22964, *Measuring and Monitoring Carbon in the Agricultural and Forestry Sectors*, by Ross W. (continued...)

too expensive for practical use. Surveying, even with current technologies, can be cumbersome and expensive, especially if forests are extensive and/or inaccessible. Geographic information systems (GIS) can facilitate gathering and organizing survey data into electronic maps, but add to the total cost of data measurement.

Remote sensing from aircraft or satellites can be used for measuring forests. Because of the extent of forests and the sometimes difficult access, remote sensing has for decades been used to map and calculate forest area, density, and other measures. What can be measured and how it is measured has changed as imaging technologies have evolved. Current imaging technologies use an array of wavelengths for developing images, including radio waves (radar) and light waves (lidar), and some technologies rely on multiple wavelengths to develop a more complete image. Imaging technologies also differ in resolution (measurement scale of the "pixels" used for recording and displaying data). High-resolution imagery can distinguish areas as small as one square meter on the ground (each pixel is thus one square meter). Moderate-resolution images are commonly 30 meters on a side, or 900 square meters (nearly a tenth of a hectare per pixel), while coarse-resolution images may be 100 meters on a side (each pixel is a hectare).

Remote Sensing Data Collection

Remote sensing has limitations—the cost of the technology to gather and use the data. Two aspects of remote sensing significantly affect the cost:

- **Remote sensing platform.** *Satellites* are multi-billion-dollar investments to develop and launch. The data they provide are quite useful, but their high investment cost necessarily means data collection and reporting for multiple purposes, of which forest measurement may be a relatively low priority. *Aircraft* can also be used, but because they fly at much lower altitudes, many more overflights of an area are needed to generate a comprehensive picture.

- **Data resolution.** Higher resolution increases the cost to construct the imaging equipment, might increase the number of overflights needed, and significantly increases the amount of data collected. At the coarse (one hectare per pixel) resolution, the world's tropical forests encompass 1.8 billion pixels; at the fine (one meter per pixel) resolution, they encompass 18 trillion pixels (with multiple data streams for each pixel).

In addition, clear conditions are required for many remote sensors, as clouds interfere with the images; sensors that can "see" through clouds exist, but are not yet widely deployed. This can be problematic for tropical rainforests, since clouds and rain are common phenomena. Thus, multiple overflights/satellite passes may be needed to generate a comprehensive picture.

Remote Sensing Data Utilization

Once the data have been collected, they must be aggregated into the comprehensive picture. This requires that pixels from adjoining overflights/satellite passes be matched to assure complete, non-duplicated coverage. Also, the data streams must be converted from the images (heights,

---

(...continued)

Gorte and Renée Johnson.

texture, infrared heat signature, etc.) into usable information on land use (e.g., intact forests, degraded forests, pastures, or cropland) and other relevant matters (e.g., biomass quantities or soil carbon levels). The data conversions (algorithms) are generally proprietary information, so users develop their own or purchase an existing conversion package. The resulting information must then be "ground-truthed"—the results for a particular site must be compared to actual forest conditions of that site to assure that the algorithm produces accurate information. Developing and ground-truthing the data conversion algorithms is, as with everything else, expensive and time-consuming, and more ground-truthing is more expensive, but increases trust in the validity of the results. Ground-truthing is also particularly problematic for remote or inaccessible forests.

The variation in data conversion algorithms is one of the sources of differences in the reported forest area data. As noted above, what is a tree and what is a forest are not always easy to define, especially at the edges of forest biomes. Thus, one might expect different algorithms to result in different forest data, even from the same remotely sensed data streams.

Significant costs can be incurred in acquiring the technology and technical expertise to generate and use remotely sourced forest data. Developing countries can be in a particularly difficult position in obtaining accurate, current forest data. They often lack both the technology and the technical expertise to generate and use remotely sensed data, and often lack the funding to acquire the technology and expertise. Developing countries may even lack the funding to acquire the results.

The results of remote sensing raise interesting issues of proprietary rights and national sovereignty. Clearly, organizations that develop and deploy the remote sensing technology and the data conversion algorithms have a financial interest in the remotely sensed data, and their sale of that information is the reward for investing time, money, and people in developing the technology. However, some question whether data about forests is public information that should be available to anyone. At one extreme, some argue that data about publicly owned resources, such as forests, should also be public, and that the owners of the resources (the public) should not be required to pay for the data. At the other extreme, some countries argue that their forests do not belong to the world, and the world has no right to information about their forests. The question: when public (e.g., U.S. or U.N.) resources are used to support data collection on forests, should those data be globally public? Would this still be true if the forest data are collected without the support or approval of the country where the forests are located? In other words, if the United States has the satellites and technology to assess another country's forests, does the United States have the right and/or the responsibility to make that information available for the public good?

Conclusion

Lowering CO_2 emissions is a central focus of U.S. and international climate change policy. An estimated 75%-80% of global CO_2 emissions stem from industrial sources, specifically burning fossil fuels. About 20% of emissions are attributed primarily to deforestation. Some contend that reducing deforestation is one of the least costly methods of reducing CO_2 emissions,[154] and that "forestry can make a significant contribution to the low-cost global mitigation portfolio."[155] One

---

[154] Kindermann et al., "Global Costs Estimates of Reducing Carbon Emissions Through Avoided Deforestation."

[155] G. J. Nabuurs et al., "Forestry," in *Intergovernmental Panel on Climate Change, Climate Change 2007: Mitigation of Climate Change*, ed. B. Metz et al. (New York, NY: Cambridge University Press, 2007), p. 543.

study found that a 10% reduction in deforestation between 2005 and 2030 could provide emissions reductions of 0.3-0.6 $GtCO_2$ per year (about 5%-10% of U.S. emissions) at a cost of $0.4 billion-$1.7 billion annually.[156]

Forests occur around the globe, at many latitudes. Many are concerned with the possible impacts of losing boreal and temperate forests. However, existing data show little, if any, net deforestation in these ecosystems, and the carbon consequences of boreal and temperate deforestation are relatively modest. In contrast, the loss of tropical rainforests is substantial and ongoing, with significant climate impacts because of the large amount of CO_2 currently stored in vegetation in the tropics—40%-50% of the carbon in all terrestrial vegetation. Thus, the largest cost and carbon benefit of reducing deforestation is with tropical forests.

Measuring forests is complicated. Definitions differ. Forests are extensive and often inaccessible. Technologies to assess forests remotely exist, but are expensive and their availability is limited. Monitoring deforestation adds to the difficulty and complexity, because forest areas must be measured repeatedly, using consistent definitions and technologies. Compensating landowners and/or countries for reducing deforestation requires that measuring and monitoring forests become more standardized. Existing forest area data, and especially the data on forest area changes, should be used with caution, perhaps seen more as indicative than as precise, accurate measurements.

The causes of tropical deforestation are manifold, and vary regionally around the globe. In some places, the drivers are commercial logging, followed by slash-and-burn agriculture that may prevent regrowth of tropical forests. Elsewhere, the major cause of deforestation is large-scale commercial agriculture, especially for cattle ranching, soybeans, and oil palm. Deforestation may also result from weak land tenure and/or weak or corrupt governance to protect the forests. Nonetheless, there is a broad consensus that the highest potential for reduced deforestation is in tropical regions where forests are abundant, carbon stocks are high, and the threat of deforestation is high.[157] Further, reducing deforestation in the tropics would likely have ancillary benefits, including preserving biodiversity, providing livelihoods for rural poor, and sustaining indigenous communities and their cultures, among other things.

Policies and practices to reduce emissions from deforestation and degradation (REDD) vary considerably and depend on several factors that are particular to the regions they address. Some forestry practices can reduce the impacts of net deforestation, and several market approaches are evolving that could compensate landowners for not deforesting their lands. Existing U.S. programs provide overseas development assistance to conserve forests, but funding levels have been modest. Also, the programs are relatively narrow in their approach to forest conservation, and in some cases, require outstanding debt to the United States to generate funding.

Some of the challenges for implementing REDD programs include accurately and effectively monitoring REDD activities and projects and improving the capacity of developing countries to implement REDD programs and to ensure compliance. Evidence from past efforts to reduce deforestation as well as from existing data on forests and deforestation suggest this might be a significant challenge. Congress is considering REDD in pending climate legislation (e.g., H.R.

---

[156] Kindermann et al., "Global Costs Estimates of Reducing Carbon Emissions Through Avoided Deforestation."

[157] E. C. Myers Madeira, *Policies to Reduce Emissions from Deforestation and Degradation (REDD) in Developing Countries*, Resources for the Future, Washington, DC, 2008.

2454 and S. 1733).[158] Also, REDD was discussed in Copenhagen in December 2009 at a Conference of the Parties to the United Nations Framework Convention on Climate Change (UNFCCC), and is expected to continue to be significant in future UNFCCC negotiations. Options being discussed include funding to improve developing country capacity (e.g., inventories to establish national baselines, training for law enforcement to combat illegal logging, and improvements in governance and land tenure systems) and mechanisms to fund national and subnational deforestation reduction activities.

---

[158] See CRS Report R40990, *International Forestry Issues in Climate Change Bills: Comparison of Provisions of S. 1733 and H.R. 2454*, by Pervaze A. Sheikh and Ross W. Gorte.

Appendix. Selected U.S. Programs That Address Deforestation

Federal Agency Activities

United States Agency for International Development

The U.S. Agency for International Development (USAID) is an independent federal agency established to administer international economic and humanitarian assistance programs, in conjunction with the Department of State. USAID has international and regional programs that address international forest conservation. In particular, the Biodiversity Program (22 U.S.C. §2151q) aims to help developing countries maintain biological diversity, wildlife habitats, and environmental services. The program funds projects and activities throughout the world, emphasizing sustainable development and community-based conservation. The program began in the 1970s to address the conservation of forests, and later expanded to address biological diversity and tropical deforestation in the 1980s.

USAID also coordinates with six U.S.-based nongovernmental organizations through the Global Conservation Program. This program was initiated in 1999 to promote landscape-scale conservation in high-priority ecosystems, where partner organizations work toward reducing conservation threats (e.g., wildlife poaching and illegal logging) and building capacity in local groups. This program is being implemented to conserve forests in Indonesia and Papua New Guinea, in the Democratic Republic of the Congo, and in Bolivia, Colombia, Ecuador, and Peru.[159]

In addition, USAID administers region-specific programs related to forest conservation, such as the Amazon Basin Conservation Initiative and U.S. participation in the Congo Basin Forest Partnership. The Amazon Initiative aims to conserve biodiversity (which includes forests) managed by indigenous and traditional groups, and to promote regional cooperation for sharing knowledge and improving governance to help conserve resources of the Amazon basin. Objectives of this program include maintaining forest cover and maximizing use of non-timber forest products (e.g., fruits and nuts). The Congo partnership is similar, but involves several additional countries. The United States financially supports the Congo Partnership through the USAID Central African Regional Program for the Environment (CARPE), which began as a regional initiative in 1995. The Congo Partnership and CARPE focus on projects to support a network of managed protected areas, to improve forest governance, and to develop sustainable management practices for resource use in the Congo basin.

U.S. Forest Service

The U.S. Forest Service (FS), within the Department of Agriculture, administers the National Forest System; conducts research on forest management, protection, and use; and provides financial and technical assistance to other forestland owners. FS has an International Program that

---

[159] This program has provided more than 40 grants totaling $110 million since its inception in 1998. See http://www.usaid.gov/our_work/environment/biodiversity/pubs/gcp_brochure.pdf for more information.

promotes sustainable international forest management and biodiversity conservation. The program supports specific activities that include managing protected areas, protecting migratory species, engaging in landscape-level forest planning, providing fire management training, curbing invasive species, preventing illegal logging, promoting forest certification, and reducing the impacts of forest use.

U.S. Department of the Interior

The U.S. Department of the Interior has two agencies that assist with global forest conservation: the National Park Service (NPS) and the U.S. Fish and Wildlife Service (FWS). NPS has an International Program that helps other nations establish and manage park systems. This program helps poorer countries benefit from conservation, cultural heritage, and recreation opportunities. NPS has provided technical assistance and training to foreign agencies that manage park systems containing forests.

FWS addresses international wildlife conservation and trade and implements relevant U.S. wildlife laws through its International Affairs office. FWS implements the Convention on International Trade in Endangered Species of Wild Flora and Fauna (CITES), to which the United States is a party.[160] CITES indirectly promotes forest conservation by regulating the trade of several tropical timber species that are listed in appendices to the agreement.

The FWS International Affairs office coordinates programs that address forest conservation indirectly by supporting the conservation of species and ecosystems. It is responsible for supporting wildlife conservation initiatives around the globe. For example, it implements the Multinational Species Conservation Fund (MSCF), supporting conservation efforts (including habitat protection) for tigers, the six species of rhinoceroses, Asian and African elephants, marine turtles, and apes (gorillas, chimpanzees, bonobos, orangutans, and the various species of gibbons). The fund provides grants to foreign countries to help build law enforcement capacity, mitigate human-animal conflicts, conserve habitat, conduct population surveys, and support public education programs. The program is active on the islands of Borneo and Sumatra, as well as in Russia, India, Indonesia, Nepal, and several African countries.

Further, FWS implements the Wildlife Without Borders Program. This program funds conservation activities through four regional initiatives: (1) Latin America and the Caribbean; (2) Mexico; (3) Russia and East Asia; and (4) Near East and South Asia. The program funds projects for training wildlife managers and conserving species of international concern, including tree species and forest habitats for animal species. These projects could also include habitat management training, education, information and technology exchange, and networks and partnerships for professionals in developing countries.

U.S. Department of State

The international conservation programs of the U.S. Department of State assist in negotiating global treaties, promoting treaty enforcement, developing international initiatives addressing sustainable development and conservation, and creating a foreign policy framework addressing

---

[160] CITES is implemented domestically through the Endangered Species Act (ESA; P.L. 93-205; 16 U.S.C. §§1531-1540). For more information, see CRS Report RL32751, *The Convention on International Trade in Endangered Species of Wild Fauna and Flora (CITES): Background and Issues*, by Pervaze A. Sheikh and M. Lynne Corn.

U.S. interests. Specifically, the Office of Ecology and Natural Resource Conservation coordinates the development of U.S. foreign policy approaches for managing ecologically and economically important ecosystems, including forests, wetlands, coral reefs, and the species that depend on these areas. The office also advances U.S. interests in a variety of international organizations, institutions, treaties, and other forums, including the United Nations Forum on Forests.

Bilateral Efforts

Debt-for-Nature Swaps Under the Tropical Forest Conservation Act

Congress enacted the Tropical Forest Conservation Act (TFCA; P.L. 105-214; 22 U.S.C. §2431) in 1998 to protect tropical rainforests for preserving biological diversity, reducing atmospheric carbon dioxide, and regulating hydrological cycles. TFCA authorizes "debt-for-nature" transactions, where developing country debt is exchanged for local conservation funds to conserve tropical forests. To be eligible, a developing country must contain at least one tropical forest with unique diversity, or a tropical forest tract that is representative of a larger tropical forest on a global, continental, or regional scale. Conservation funds (in local currency) from these exchanges are deposited in a tropical forest fund for each country. Interest earned from the principal balance, as well as the principal itself, is usually given as grants to fund tropical forest conservation projects. Eligible conservation projects include (1) establishing, maintaining, and restoring forest parks, protected reserves, and natural areas, as well as the plant and animal life within them; (2) training to increase the capacity of personnel to manage reserves; (3) developing and supporting communities near or within tropical forests; (4) developing sustainable ecosystem and land management systems; and (5) identifying the medicinal uses of tropical forest plants and their products.

Free Trade Agreements

The United States has developed free trade agreements (FTAs) with many countries, and is negotiating FTAs with other countries. Some of the negotiations have addressed illegal logging, which is a significant contributor to tropical deforestation in some areas. For example, in 2006, the United States and Indonesia signed a memorandum of understanding (MOU) to enhance bilateral efforts to combat illegal logging and associated trade. The United States committed $1 million with this agreement to fund projects that would reduce illegal logging in Indonesia, such as using remote sensing to identify illegally logged tracts of land. The MOU also set up a working group to assist in implementing the initiative under a pending U.S.-Indonesia Trade and Investment Framework Agreement.

Similarly, a third-party agreement within the U.S.-Peru FTA is expected to increase awareness of illegal logging in Peru and add additional mechanisms to address illegal logging. The third-party agreement requires each country to effectively enforce its own environmental laws that affect trade between the parties. Further, it establishes a policy mechanism to address public complaints that a party is not effectively enforcing its environmental laws, regardless of whether the failure is trade-related.

U.S. Involvement in International Programs

Global Environmental Facility

The Global Environment Facility (GEF) was established in 1991 to fund international environmental needs in four areas: climate change, stratospheric ozone depletion, biological diversity, and international waters. In recent years, GEF has also addressed land degradation—particularly deforestation and desertification—and persistent organic pollutants. The GEF is designed to provide incremental funding to cover additional costs for development projects needed to provide environmental benefits connected to issues on the GEF agenda. However, GEF has evolved into funding a variety of activities for planning, including national action plans, in addition to providing incremental funding for specific projects. Some 176 donor and recipient nations, including the United States, are participants in GEF, and meet every four years in a General Assembly to agree on funding levels.

International Tropical Timber Organization

The International Tropical Timber Organization (ITTO) was founded in 1986 under the auspices of the United Nations because of concerns over tropical deforestation. The organization was derived from the International Tropical Timber Agreement,[161] which provides a framework for tropical timber producing and consuming countries to consult on issues related to international trade of tropical timber, and methods of improving forest management to promote conservation. The ITTO has 60 members (including the United States), which together have about 80% of the world's tropical forests and conduct 90% of the global tropical timber trade. The ITTO promotes sustainable forest management and forest conservation strategies, and assists tropical member countries in adopting such strategies in timber harvesting projects. The ITTO also collects, analyzes, and disseminates data on the production and trade of tropical timber.

United Nations Reduction in Deforestation and Forest Degradation Program (UNREDD)

The United States provides expertise to the United Nations and other countries for developing global forest carbon accounting systems through UNREDD and advancing carbon markets. FS experts in forest inventory and monitoring technology and in carbon cycle modeling have been working with the FAO to develop carbon accounting methods for forests worldwide. These experts are helping international communities determine how governments can be paid for the service of carbon sequestration in forests. Other U.S. agencies are working with the FS to track forest cover worldwide. In the Department of the Interior, the U.S. Geological Survey's data acquisition platform, Landsat, provides remotely sensed data that are interpreted using FS forest inventory and monitoring information. This information can be used by UNREDD in worldwide applications.

---

[161] See http://www.itto.int/en/itta/.

Author Contact Information

Ross W. Gorte
Specialist in Natural Resources Policy
rgorte@crs.loc.gov, 7-7266

Pervaze A. Sheikh
Specialist in Natural Resources Policy
psheikh@crs.loc.gov, 7-6070

www.ingramcontent.com/pod-product-compliance
Lightning Source LLC
Chambersburg PA
CBHW081357170526
45166CB00010B/3119